Further Discoveries about the Surveying and Planning of Roman Roads in Northern Britain

A sequel to BAR 492

John Poulter

BAR British Series 598
2014

Published in 2016 by
BAR Publishing, Oxford

BAR British Series 598

Further Discoveries about the Surveying and Planning of Roman Roads in Northern Britain

ISBN 978 1 4073 1281 1

BAR Publishing is the trading name of British Archaeological Reports (Oxford) Ltd. British Archaeological Reports was first incorporated in 1974 to publish the BAR Series, International and British. In 1992 Hadrian Books Ltd became part of the BAR group. This volume was originally published by Archaeopress in conjunction with British Archaeological Reports (Oxford) Ltd / Hadrian Books Ltd, the Series principal publisher, in 2014. This present volume is published by BAR Publishing, 2016.

Printed in England

BAR
PUBLISHING

BAR titles are available from:

BAR Publishing
122 Banbury Rd, Oxford, OX2 7BP, UK
EMAIL info@barpublishing.com
PHONE +44 (0)1865 310431
FAX +44 (0)1865 316916
www.barpublishing.com

Contents

List of figures

Note: nearly all of the figures in this monograph are line diagrams, in which, as a rule, the planning alignments are shown as dashed lines, and the Roman road (as built) as a solid line. However, where the course of the Roman road may not be clear, or not central to the argument, it is sometimes shown as a dotted line or one with short dashes. In general the following symbols have been used in the diagrams to indicate:

\+ the location of a place, such as a fort, town, or village; although on a few figures ■ has been used to indicate specifically the location of a Roman fort

▲ the location of a significant hill

➤ the direction of flow of rivers and streams

➞ the probable direction of planning.

Introduction and acknowledgements

The research described in this monograph follows on directly from those findings which were reported in my two previous archaeological publications. These were Surveying Roman Military Landscapes across Northern Britain, which was published as BAR monograph number 492 by BAR Publishing in 2009, and The Planning of Roman Roads and Walls in Northern Britain, which was published by Amberley in 2010.

Although these publications had reported significant discoveries about the planning of Roman roads and walls, they had left a number of questions outstanding. This included a query about what might have been the purpose or purposes of the long-distance alignments which had been found to underlie many stretches of Roman Dere Street and Hadrian's Wall. It was asked: might these alignments have been more than simply devices to help set out the courses of the road and the wall?

Despite having posed the question, I had had no particular intention of attempting to answer it myself. I felt it had become timely to draw a line under the research that I had been conducting, and that other scholars, with a wider knowledge of Roman planning and administration, might be prompted to address the question instead. However, on the very next day after the Amberley book had been published, Professor David Breeze put me in touch with Rob Entwistle, who appeared to have detected, quite independently, several more long-distance Roman alignments, particularly in the north of England.

This considerable expansion of the body of evidence for such alignments stimulated renewed interest in the question and provoked an extended correspondence between Rob Entwistle and myself. In the course of this correspondence several possible explanations for the creation of these alignments were constructed, assessed, refined and, in many cases, rejected. This monograph is built upon those possible explanations which were not rejected, and I am grateful to be able to draw upon some of Rob Entwistle's findings – as well as further discoveries of my own – to help illustrate why these are considered to be feasible.

However, my archaeological focus is a relatively narrow one. It rests upon the planning of Roman roads (and walls), and the mechanisms by which their courses may have been set out on the ground. From an overall perspective, it is possible that the nature of Roman planning across newly-acquired territories will prove to have been much more all-embracing and involved than simply deciding upon the locations of their forts and the lines of their roads. Therefore the full publication of Rob Entwistle's observations and analyses is awaited with considerable interest.

In the meantime, during the aforementioned correspondence, contact was also made with Dr John Peterson, who is perhaps best known for his work on Roman centuriation and the possibility of its existence in Britain. The practice of centuriation - which is described in literature from Roman times - is the means by which certain quite large areas of countryside were surveyed and divided up by the Romans into regular squares or rectangles. This is perceived as having been undertaken for the purposes of allocating parcels of land to various individuals, such as, conceivably, veterans from the Roman legions. With regard to this monograph, the key point that John Peterson has made is that the practice of centuriation might have been employed not just for determining land ownership but also for determining the locations of forts and settlements and the roads that would run between them. As John Peterson has remarked, the creation of a centuriated landscape implies the creation of maps, and if you have a map you may as well make use of it.

The upshot of all the foregoing is that some of the proposals in this monograph are radical. However, they have all been derived purely from the evidence on the ground and after very careful analysis, which, I believe, has been as objective and as scrupulous as it is possible to be. Hence they are submitted for serious consideration by those scholars who are interested in the planning and surveying of the networks of Roman roads and military installations, and particularly those which lie in northern England.

I am very grateful to Dr David Davison of BAR Publishing for being willing to consider publication of this monograph. Initially, this had been simply on the basis of a synopsis, but this agreement has allowed me from the outset to design and create the layout of the text and diagrams to suit the BAR style.

I am very grateful, too, to David Breeze for agreeing to read the draft of the text and for raising many pertinent questions about the relationship of these findings to the main body of knowledge about Roman surveying. I am also very grateful to Professor Brian Campbell for reviewing the draft and making many suggestions with regard to the context of Roman surveying practice. Both sets of feedback have enabled the content of this final version to be considerably enriched with additional information and sources of reference.

I am very grateful as well to Paul Bidwell for reading and commenting upon the draft text. This has assisted me in making several points more clearly and rationally for the benefit of readers in general. In addition, quite apart from drawing upon some of his findings, I have to thank Rob Entwistle particularly for making many comments and suggestions upon the draft text – and upon some of the revisions, too. These have, again, helped greatly with the clarification of the points which I have been trying to make with regard to the planning of Roman roads.

Very helpful comments, suggestions and many valuable elements of information have also been received from Mike Haken, Dr Nick Hodgson, Dr David Mason, John Peterson, Dave Robson, High Toller and Humphrey Welfare. Their inputs are all acknowledged with gratitude. Finally, thanks are expressed to my friend Barbara Tearle, who kindly spent several days running her expert editorial eye over the draft text and diagrams. Numerous errors and inconsistencies were identified, the eradication of which has resulted in a much neater and better-presented document as a whole.

Ultimately, though, for better or for worse, the responsibility for the content of this monograph rests with myself. It is hoped that both Roman scholars and general readers may find it illuminating and a worthwhile contribution to the understanding of how the Romans may have planned the networks of roads and military installations which still reflect such an imposing pattern of regulation and control across the landscapes of northern England.

John Poulter
Wing, Bucks.
March, 2014

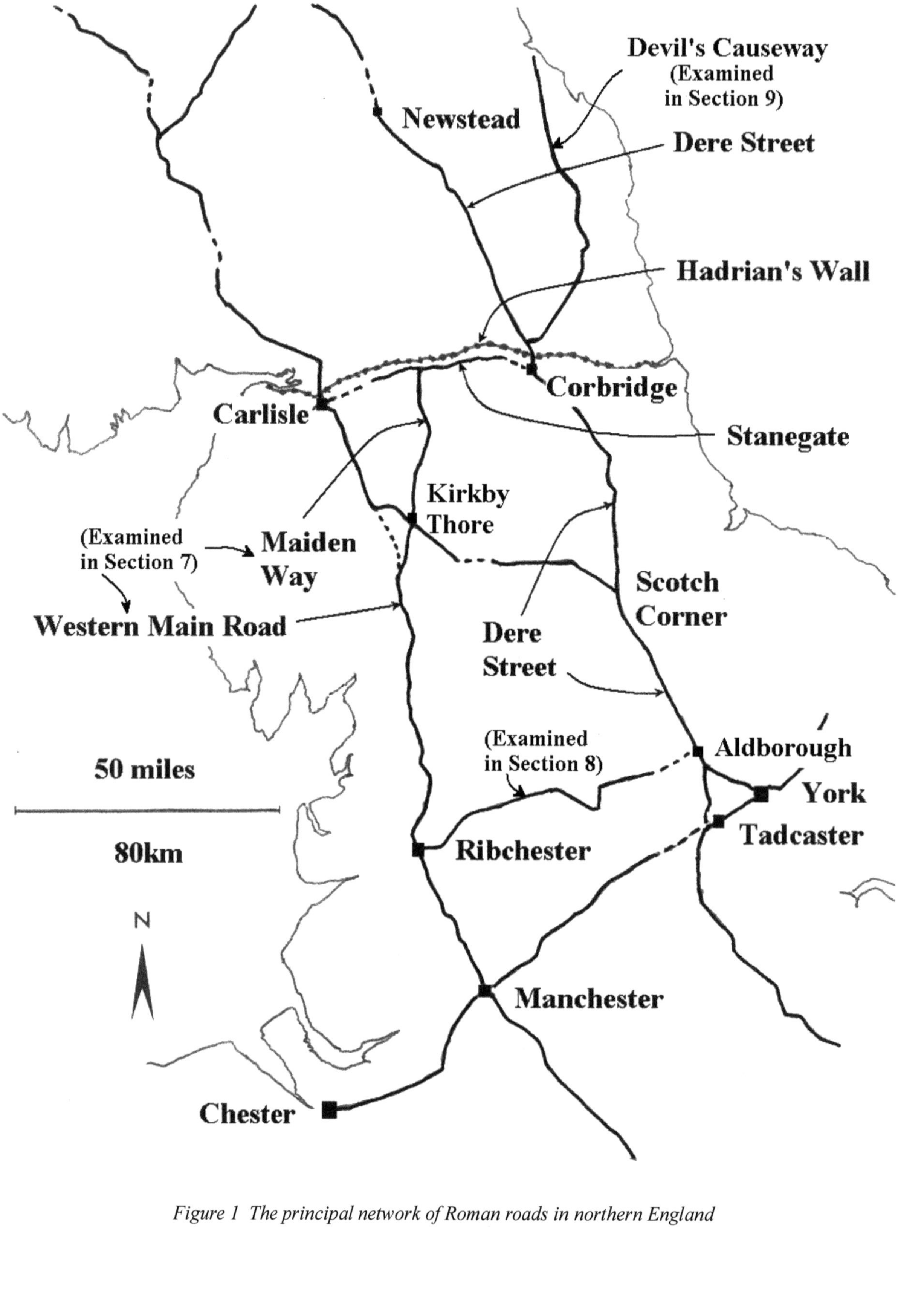

Figure 1 The principal network of Roman roads in northern England

Section 1: A recapitulation of previous work

This monograph aims to address a question which was posed in both of the author's former publications about the planning of Roman roads in northern Britain.[1], [2] Among the discoveries reported in those publications was the fact that, in many cases, the setting-out of a Roman road appeared to comprise a two-stage process. In the first stage, it appears that long-distance alignments were laid out absolutely straight across the landscape, often for distances of 20 to 30 miles (32 to 48 km), and sometimes more. Then, probably when the time came to build the road, it appears that deviations were set out from these long-distance alignments where necessary to meet local objectives, such as to service a fort or a settlement away from the long-distance alignment, or to reach a convenient river crossing, or to avoid unsuitable ground.

This discovery was made whilst examining the course of Roman Dere Street from the Vale of York up to Newstead in Scotland. Subsequently, the same two-stage process was also found to have been applied to the setting-out of the line of Hadrian's Wall. Hence this process appears to have been a fairly standard Roman surveying technique at the time.

However, from an examination of the archaeological evidence, there are good grounds for suspecting that, along Dere Street, certain of the long-distance alignments may have been set out as many as 30 years before some parts of that Roman road came to be constructed. In addition, at least one other Roman road – the Rudgate – appeared to have made use of one of the long-distance alignments that underpinned the course of Dere Street. Moreover, at least two Roman forts, and possibly a third, were found to lie exactly on these long-distance alignments, even though Dere Street, as built, had deviated slightly from the alignments at these places. The question that was posed, therefore, was this: might these long-distance alignments have been created originally for some other purpose, and then been re-used, in part, when the time came for the Romans to lay out the permanent courses of their roads, such as Dere Street and the Rudgate?

In seeking to address this question, this monograph examines a number of some newly-discovered long-distance Roman alignments in the north of England. Four of these had been identified, along with a number of others, by Mr Rob Entwistle, whilst working quite independently of the author, and it is to be hoped that an opportunity will soon be found for the full account of Rob Entwistle's discoveries to be published. Reference is also made to the work of Dr John Peterson on the Roman practice of centuriation, in which it is known (from both Roman literature as well as archaeological evidence) that large areas of countryside were sometimes marked out in grids of squares or rectangles for the purposes of land planning and allotment.

In addition to addressing the foregoing question, this monograph also seeks to examine the issues which an engineer or surveyor might have to face when endeavouring to set out a road so as to follow a long-distance alignment. The small divergences from long-distance alignments which can be detected in many Roman roads in Britain are an indication that the process may not always have been a straightforward one.

With regard to the author's former publications about the planning of Roman roads, it is the BAR 492 monograph, *Surveying Roman Military Landscapes across Northern Britain*, published in 2009, which is the full archaeological report of the author's previous discoveries. It contains details of all the author's observations and analyses which were carried out along his examinations of the courses of Roman Dere Street, Hadrian's Wall, and the Antonine Wall in Scotland. It is the volume to be used by those who might wish to follow the author's observations on the ground and check for themselves the interpretations placed upon them.

The Amberley book, *The Planning of Roman Roads and Walls in Northern Britain*, published in 2010, presents and discusses these same findings and interpretations in more general terms. It also describes and illustrates the use of systems models which were employed to analyse the archaeological evidence. Although written for a more general readership, the Amberley book was intended, no less than BAR 492, to offer a serious contribution to archaeological research and scholarship.

Those readers who wish to examine the full scope of the author's previous work will therefore need to consult both of these two publications. For convenience, however, and in the interests of continuity, this first section will now offer an expanded summary of some of the key points which emanated from the earlier work and which are relevant to the new findings that are reported and analysed in the sections which follow.

To begin with, the author had not harboured any ambition to examine the planning of Roman roads as such. The earlier work had started out simply as a quest to try to detect and demonstrate the direction in which the course of a particular stretch of Roman Dere Street had been planned and marked out on the ground. It will be apparent

[1] Poulter, 2009
[2] Poulter, 2010

that such planning and marking-out would have been one of the final steps required before construction of any road could commence.

In order to address the particular question which the author had in mind, a methodology was developed for diagnosing the directions in which Roman surveyors were likely to have been working when setting out the lines of their roads. To a large extent, the methodology depended upon suppositions about how the Roman surveyors would have behaved when seeking to change the course of their roads from one straight alignment to another, when marking out the course on the ground.

To illustrate the methodology, two examples may suffice.

Firstly, consider the situation in which a Roman surveyor engaged in setting out the line of the road on the ground should wish to change direction from one straight alignment to another. There are numerous examples from across Britain which show that he would have tended to do this at a high point in the landscape. However, from the highest point in the landscape it is quite likely that, in the new direction, he would only have been able to see into the far distance. Therefore, as indicated in Figure 2, it was hypothesised that in practice the Roman surveyor would have tended to continue along his existing alignment over the top of the high ground until he should come to a convenient brow, from which he could obtain a good view not only into the far distance but also over the foreground and the middle distance as well. This would have enabled him to be more assured that he would have been setting out a practical course for the road. If this supposition about his behaviour should be correct, then it can be anticipated that at the turning-point of the Roman road, there would be a good view in one direction and only an indifferent view – i.e. back up to the top of the hill behind – in the other. It follows that at the point where the change of alignment occurred, the prospect with the better view would thus indicate the direction in which the planner of the Roman road had been working, when setting out the course of the road on the ground.

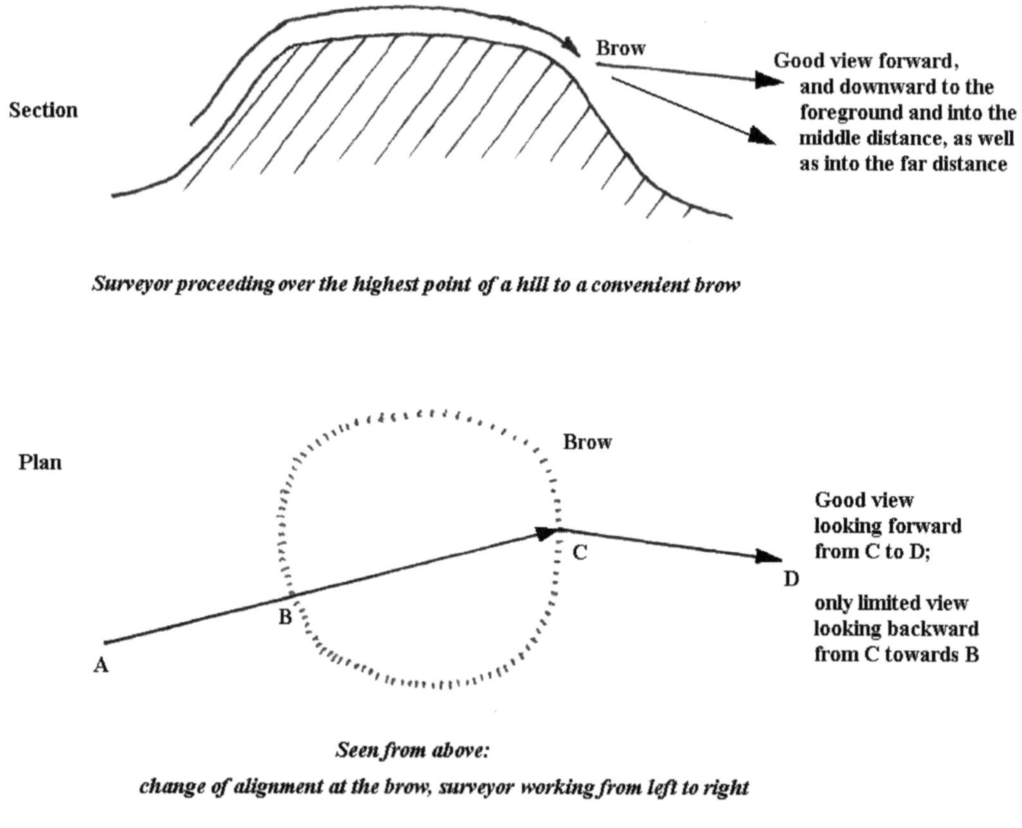

Figure 2 Changing the alignment of a Roman road at a convenient brow

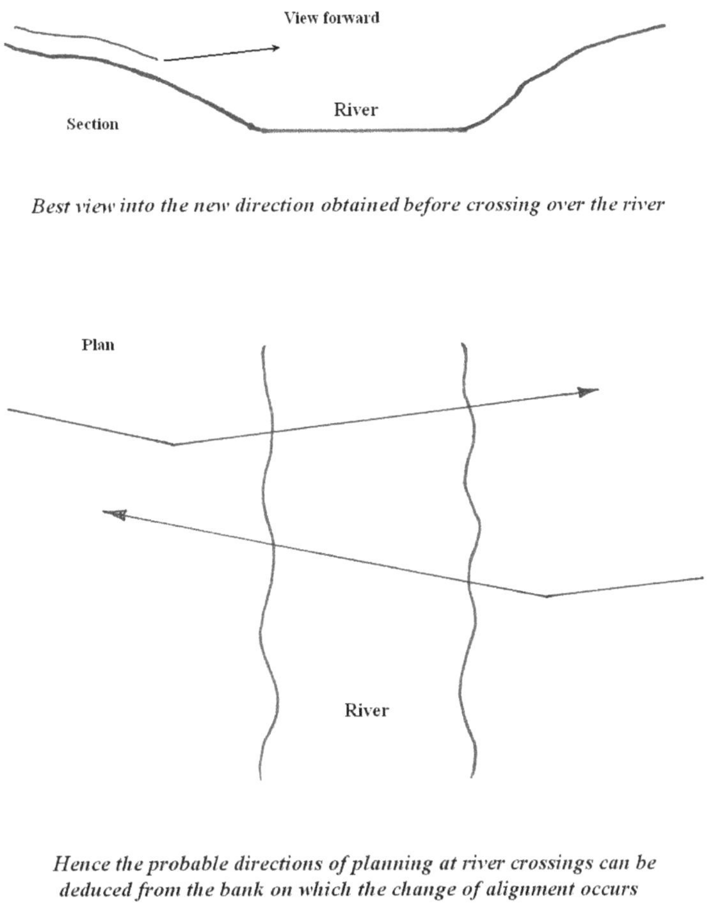

Figure 3 Changing the alignment of a Roman road at a river crossing

For a second example, consider the case of a Roman surveyor wanting to change the alignment of his road at a river crossing. This is something that did occur from time to time along Roman roads, though less frequently than on high ground. In this situation, as illustrated in Figure 3, it can be anticipated that the surveyor would have been able to obtain a better view into the new direction – and into the far distance, especially – if he had examined it from the river bank before the crossing rather than after it. Once the surveyor had crossed the river, the view into the new direction would, typically, have been largely upwards and short-ranged at best. Hence, in general, in order to be as assured as possible about the practicality of the new line for his road, it can be anticipated that the surveyor would have wanted to set out the new alignment from the bank *before* crossing the river, not *after* it. In such cases, therefore, it can be deduced that the direction in which the planner of the Roman road is most likely to have been working when setting out its course would have been from the point of turn towards the river, as shown in Figure 3.

It will be apparent that these and various related indicators of the direction of planning can only be used where the Roman road happened to have been laid out in straight lengths, with changes of direction occurring at single points of turn in between. As it happens, though, this is how many Roman roads were set out in this country. For instance, much of the course of Dere Street appears to have been laid out in this way. Accordingly, the predictions derived from these and other suppositions were tested out by visiting and examining the location of every change of direction of the aligned parts of Dere Street from the Vale of York to the Roman fort at Newstead near Melrose in Scotland. This entailed repeated inspections of more than 120 miles (192 km) of this Roman road.

To a very large extent, the results of this fieldwork appeared to confirm that the suppositions about the behaviour of Roman planners had been correct. The planners of Dere Street did indeed appear to have behaved as predicted – and to have done so consistently. Subsequently, as described below, the same methodology was applied by the author to the line of Hadrian's Wall, with results which were equally consistent and convincing.

Although the consistency of the results appeared to confirm that the author's suppositions had been on the right lines, when the time came to amalgamate the results along the total course of Dere Street, the overall picture that was presented of the nature of the Roman planning was that of organisational incoherence. Adjacent stretches of consistently-planned road appeared to have been set out in opposite directions, seemingly for no obvious reasons. A plan of the kind of results which had emerged from the fieldwork is shown schematically in Figure 4.

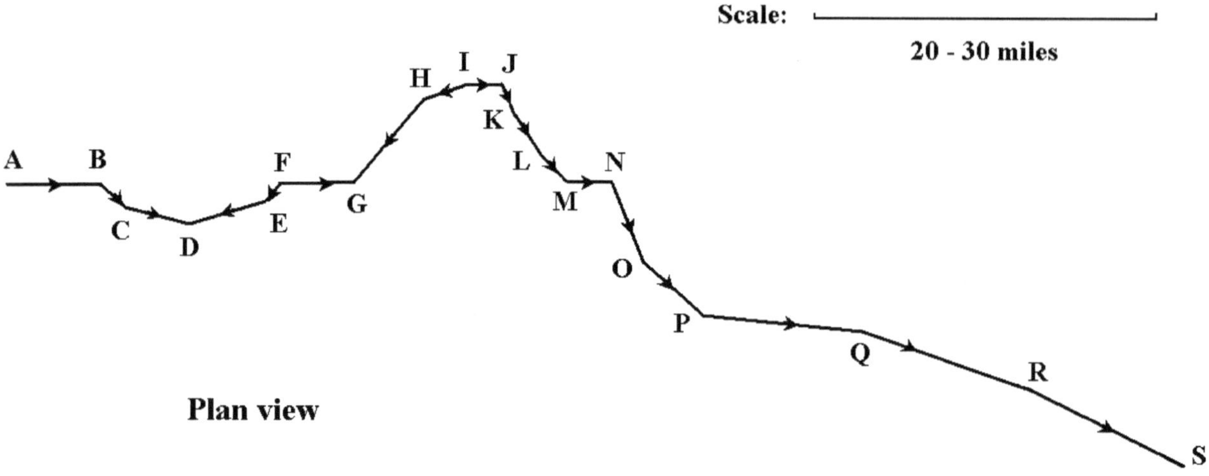

Figure 4 Directions of planning of a Roman road, shown schematically

It was at this point that the author became drawn into trying to understand the process or processes by which Roman roads might have been planned, simply because this apparently chaotic picture did not seem credible. Hence, various systems models were constructed to try to decipher the cause or causes for these apparently random changes in the directions of planning. For instance, tests were carried out to see if the Roman surveyors might have been working downhill from particular high points in the landscape, and the high points themselves were examined to see if they might in some way have been survivors from an initial long-distance surveying exercise. Alternatively, a model was constructed to explore if the work of the planners might have been partitioned between different Roman legions, as appears to have been the case with the construction of Hadrian's Wall, and with the Antonine Wall in Scotland.

Several weeks were spent in this way, trying to make sense of this apparent organisational disarray, but without success. What might appear to be a plausible possibility along one stretch of the road failed to fit along others. Then, suddenly, a particular feature was spotted, after which all the results fell into place, not just at one point but along the entire course of Dere Street. What was spotted was that, with reference to Figure 5, the stretch of road from A to B was exactly in line with stretches F to G, and M to N, and that the stretch of road from R to S was exactly in line with the change of alignment at N. In practice, these stretches could be several miles apart: it must be appreciated that, as indicated on Figure 5, the scale of the planning was a large one.

What became evident from this discovery was that the Roman surveyors appeared to have begun their planning

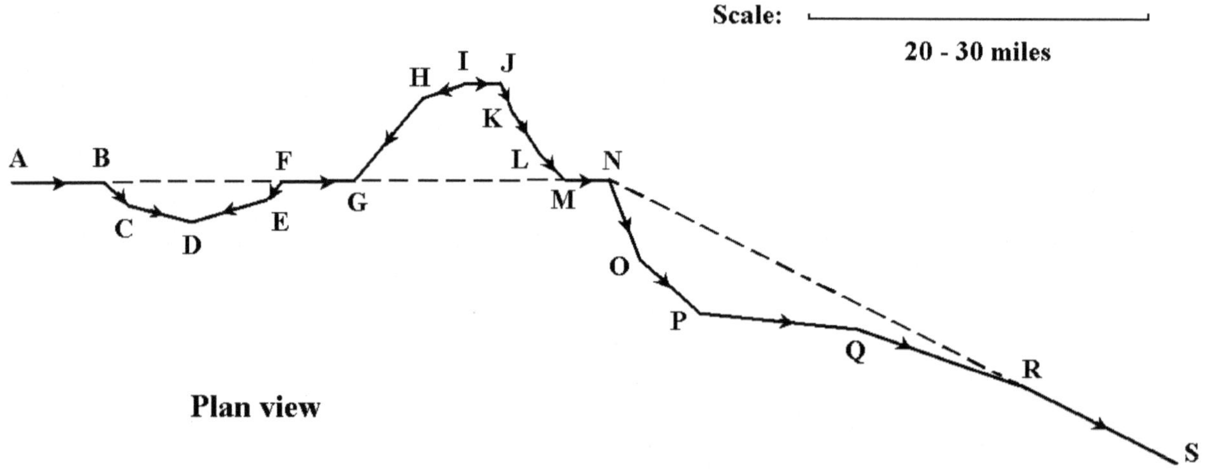

Figure 5 Directions of planning of a Roman road, showing the underlying long-distance alignments

– of at least some parts of Dere Street – by setting out a framework of long-distance alignments. These not only extended over many miles but also over multiple horizons across the landscape. Later, probably when the time came to construct the road, deviations were then set out from these long-distance alignments in order to meet local objectives, and then return to the original alignment afterwards. The local objectives might, as indicated above, have been so as to avoid unacceptably low-lying or marshy ground, or to reach a suitable crossing-point for a river, or to service a settlement or fort that might have been established at some distance from the long-distance alignments.

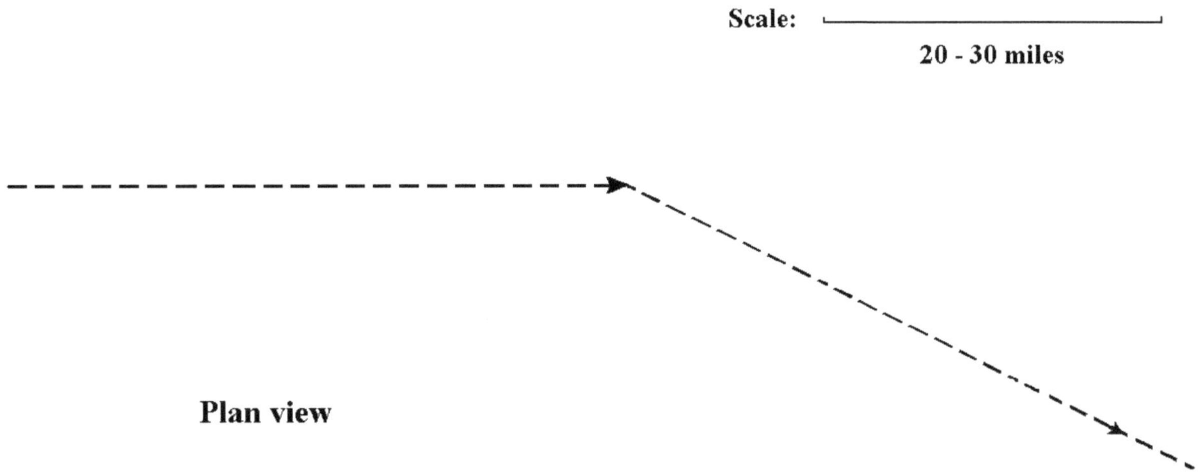

Figure 6 Process of planning the line of a Roman road: Stage 1: setting out the long-distance alignments

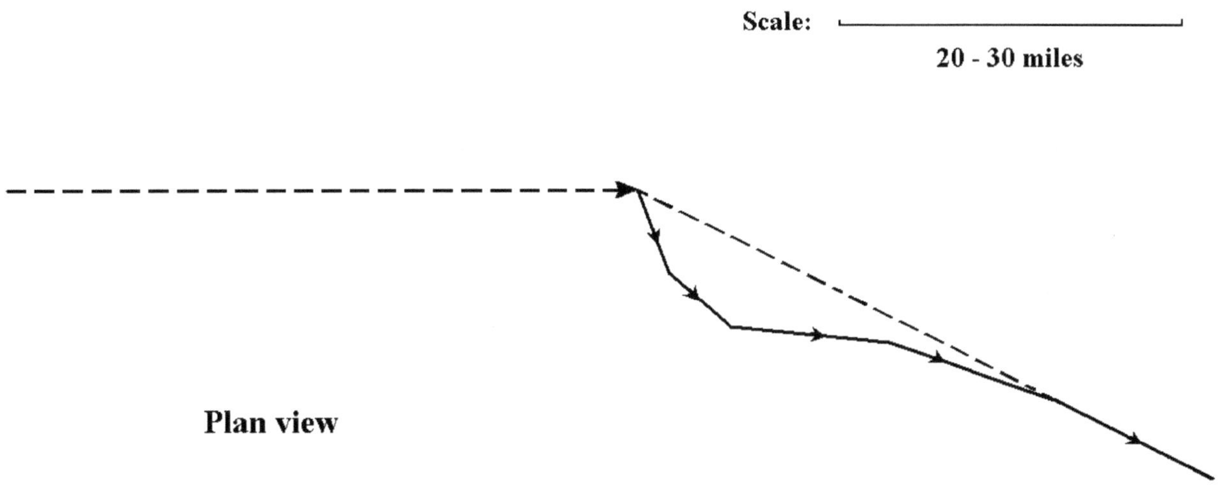

Figure 7 Process of planning the line of a Roman road: Stage 2: setting out a deviation, possibly to avoid an ill-situated valley prone to widespread flooding. In this case the deviation might have been planned in the same direction as the long-distance alignments

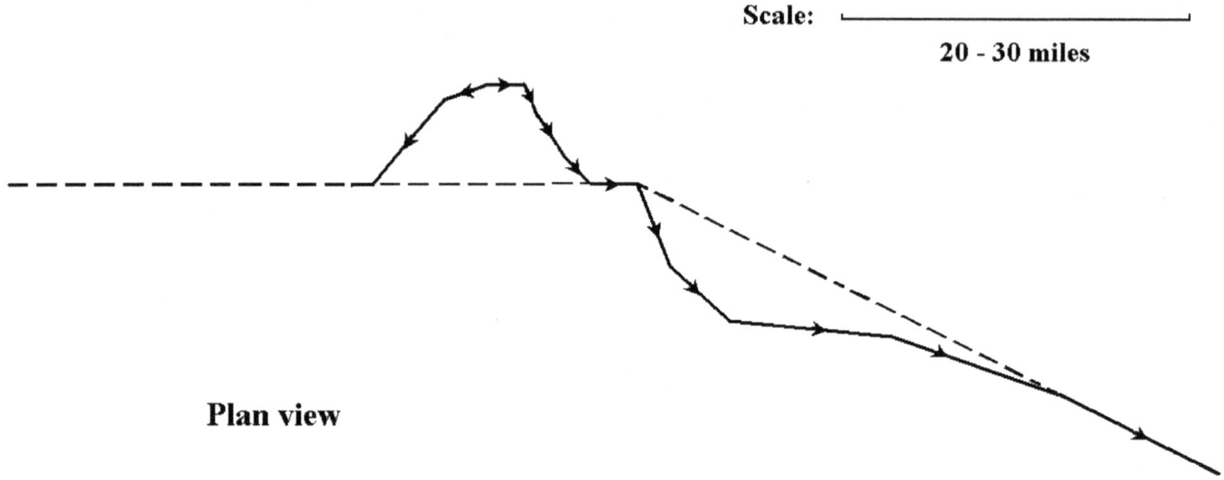

Figure 8 Process of planning the line of a Roman road: Stage 3: setting out a deviation, perhaps to cross a major river at a suitable location. In this case the deviation might have been set out from the crossing-point back to the long-distance alignment

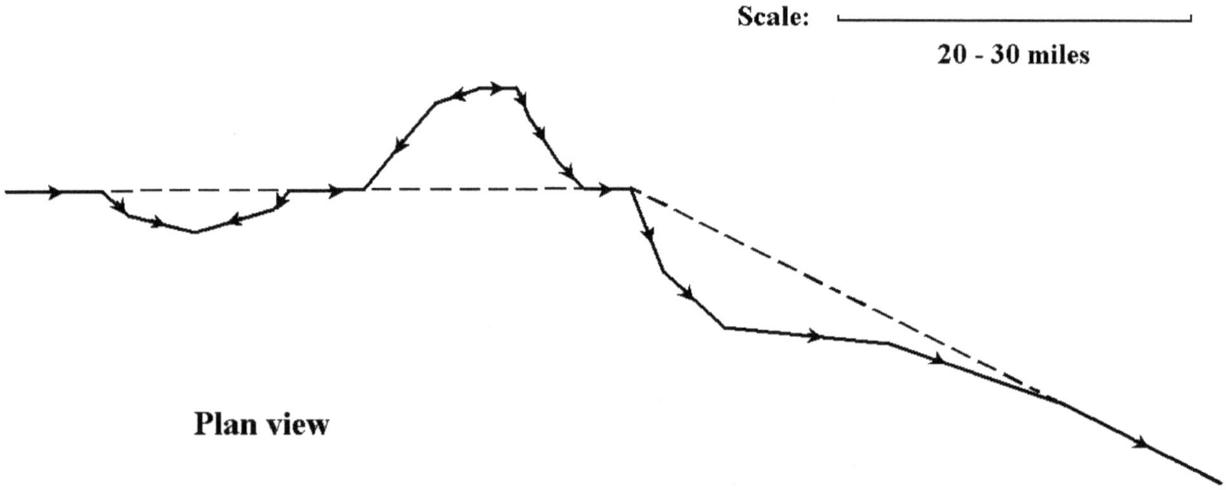

Figure 9 Process of planning the line of a Roman road: Stage 4: setting out a deviation, possibly to service a fort that had been established at some distance from the long-distance alignment. In this case the deviation might have been set out from the long-distance alignment down to the location of the fort

Figures 6 to 9 illustrate this road planning process schematically. As can be seen from these figures, the directions in which the deviations were set out seems to have been arbitrary. It appears to have been dependent upon the circumstances at the time, and was clearly not related to the directions in which the long-distance alignments had been set out.

The point which must be stressed here is that this discovery was genuinely a surprise. It was made with no preconceptions, and with data that had been assembled for another purpose – simply that of diagnosing the directions in which the courses of the Roman roads may have been set out. Before the work had started there had been no expectation that there might have been a two-stage process to the planning, with very long-distance alignments being set out first in order to underlie the courses of the Roman roads that remain to be seen on the maps today.

Once the phenomenon had been recognised, instances of its application began to be recognised in several places, as illustrated by Figures 10 to 13.

Figure 10 shows where the phenomenon was first detected. It can be seen that if the long-distance alignment of Dere Street from Healam Bridge to Scotch Corner had been followed unswervingly, then it would have taken the road through some 2 to 2½ miles (c.4 km) of the flood plain of the River Swale. This would have rendered the road liable to damage of its surface and bridges by flooding in each winter and after snow-melt in spring. Instead, as can be seen, it appears that the Romans set out a deviation for the road southwards from Scotch Corner so as to cross the Swale about 1 mile (c.1.6 km) upstream, at Catterick Bridge, and then return to the long-distance alignment at Bedale Beck, by Leeming. In this case, the long-distance alignment was

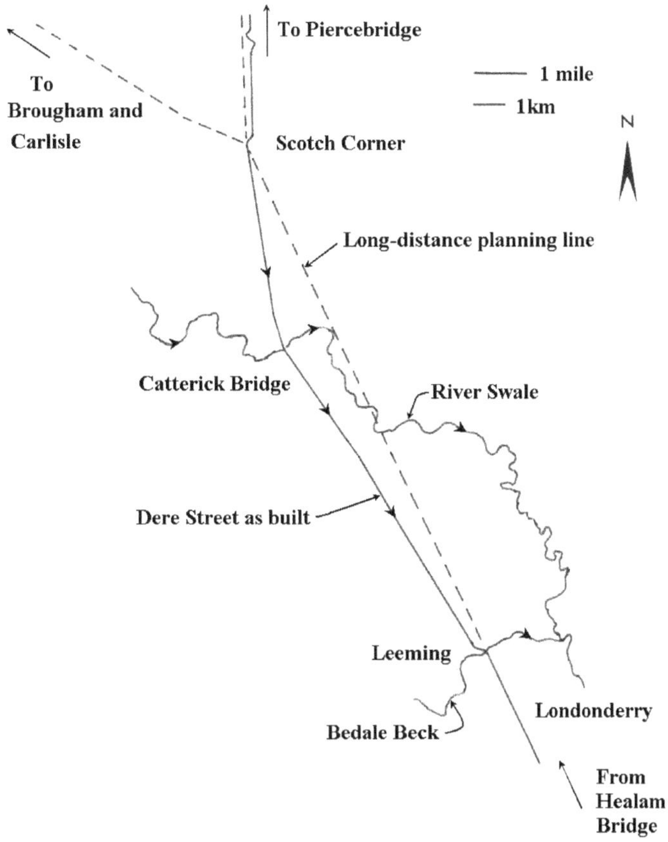

Figure 10 Deviation of Dere Street to cross the River Swale at Catterick Bridge

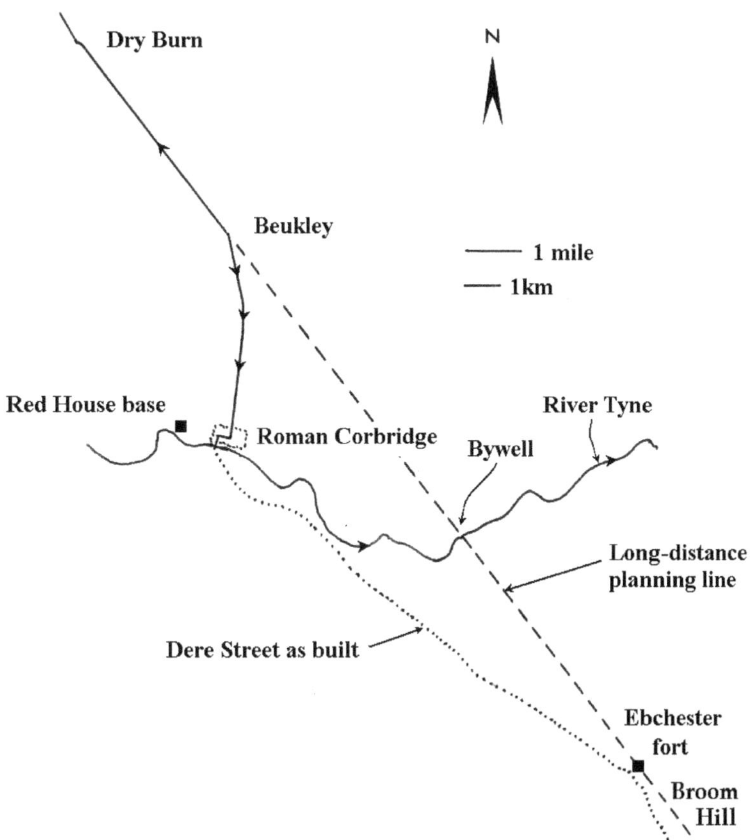

Figure 11 Deviation of Dere Street to cross the River Tyne at Corbridge rather than at Bywell

some 15 miles (24 km) in length, whilst the distance along the deviation amounted to some 11 miles (c.18 km) in all.

In Figure 11, it can be seen that a deviation for Dere Street appears to have been set out from a long-distance alignment so that the road, as built, should cross the River Tyne at Corbridge. This long-distance alignment possessed a total length of some 26 miles (42 km), and appears to have run from a village called Esh in County Durham to a minor stream some 4 miles (6.4 km) north of the River Tyne, called the Dry Burn. This alignment would have crossed the River Tyne at Bywell, but, presumably, when the time came to build Dere Street in this area, the crossing-point at Corbridge would have been the location which was preferred by the Romans. Hence the deviation, which extended for some 14 miles (22.5 km) in length, appears to have been set out from Beukley to run south to Corbridge and then, after crossing the Tyne, to return to the long-distance alignment at Broom Hill, just above the Roman fort at Ebchester.

Figure 12 shows a deviation of Dere Street that was made well to the north of the River Tyne, to service the fort at Risingham, which stood beside the River Rede. In this case the deviation was only a short one, but its return northwards to the long-distance alignment was complicated by a second deviation which had evidently been adopted to improve the sighting of the line in both directions from near Dyke Head farm.

Finally, and well away from Dere Street, Figure 13 shows a deviation running from a long-distance alignment of the Fosse Way, south-west of Leicester. This deviation had evidently been set out so as to cross the Roman road known as Watling Street at High Cross, and then return to the original alignment near Brinklow in Warwickshire. The author is indebted to Rob Entwistle for drawing this deviation to his attention, and several more of Rob Entwistle's discoveries will be described in the next section. In this case, the length of the deviation from south-west of Leicester to Brinklow amounts to some 26 miles (42 km), whilst the long-distance alignment appears to be at least 38 miles (61 km) in length.

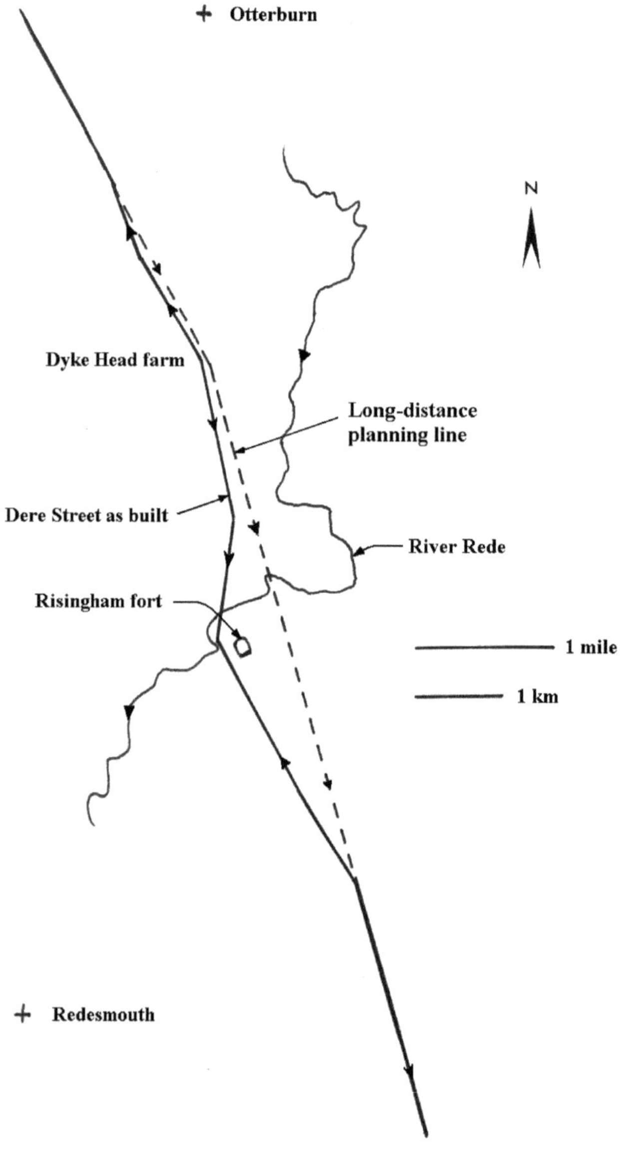

Figure 12 Deviation of Dere Street to service the Roman fort at Risingham, beside the River Rede

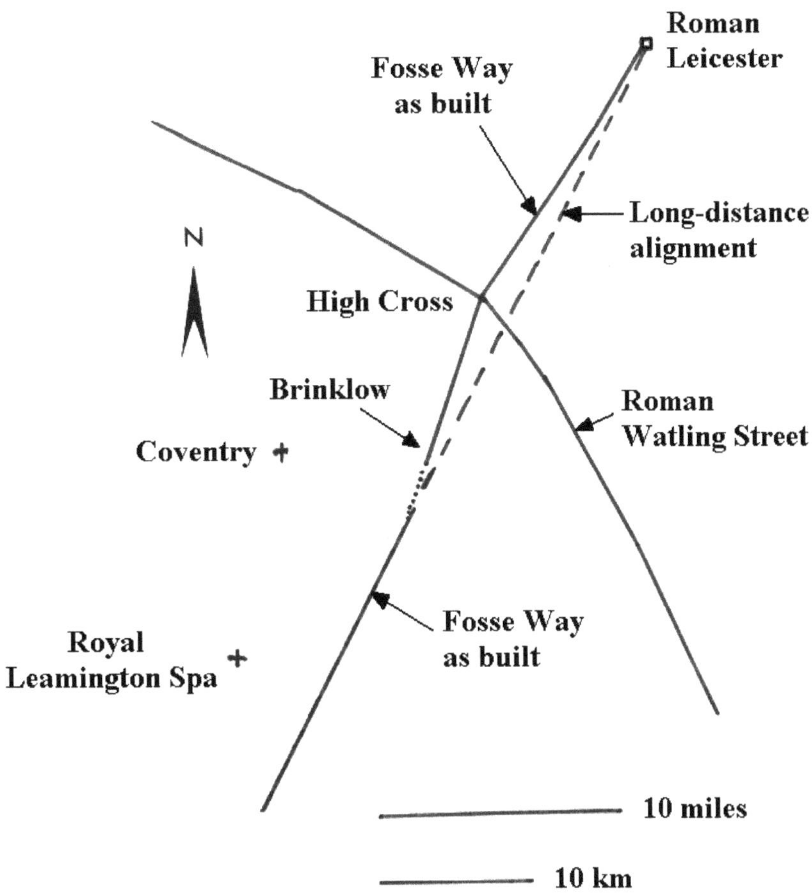

Figure 13 Deviation of the Fosse Way south-west of Leicester in order to cross Roman Watling Street at High Cross and then return to the long-distance alignment near Brinklow

As already mentioned, after the survey of the planning of Roman Dere Street had been carried out, the author then applied his methodology to the setting-out of Hadrian's Wall. It was Professor David Breeze who first suggested that the Wall should be examined in this way, and this suggestion was subsequently endorsed by Dr Nick Hodgson. Somewhat to the author's surprise, when the study was undertaken, it was found that the course of Hadrian's Wall appeared to have been laid out by the same sort of mix of long- and short-distance alignments that had been detected along Dere Street.

With Hadrian's Wall, however, the long-distance alignments were found to be line-of-sight only. That is, they did not extend beyond visible horizons, unlike several of the road-planning alignments that had apparently been used along Dere Street. Since, in the case of Hadrian's Wall, both the long-distance alignments and the shorter deviations were evidently being used to position the Wall precisely within the visible landscape, in order for it to perform its design objectives as effectively as possible, this is only to be expected.

In this case it may also be surmised that, along Hadrian's Wall, the setting-out of the long-distance alignments and the deviations from them would have been part of the same planning exercise. This is because, presumably, construction of the Wall would have commenced as soon as its line could be fully marked out on the ground. This is in contrast to the exercises in long- and short-distance surveying that seem to have determined the course of Dere Street. It appears that, in the case of Dere Street, these may have been undertaken several years apart. Nevertheless the example of Hadrian's Wall serves to demonstrate the important point that the tactic of laying out long-distance alignments and then making deviations from them was, at the time, a standard Roman practice for laying out the lines of their large linear structures such as roads and walls. In other words, it was not solely a consequence of a two-stage planning process in which the two stages had simply been separated by a period in time.

In contrast to Hadrian's Wall, when the author came to examine the planning of the Antonine Wall in Scotland, little or no use of long-distance alignments could be detected, either along or underpinning its line. By this time, the possible existence of such alignments was something that was decidedly being sought, as part of the author's research. Instead, rather than having been guided by long-distance alignments, the course of the Antonine Wall appeared to have been set out initially as a chain of forts, fortlets, and possibly signal towers, all positioned so that each of them would have been intervisible with its neighbours. The fact that these military installations appear to have been part of the initial plan for the Antonine Wall, rather than, in part, an afterthought as on Hadrian's Wall, may account for the difference in the methods of planning of the two walls, but the effect of it in Scotland would have given the Romans a continuous line of communication between the Firth of Forth and the River Clyde. The exact line of the Wall on the ground then appeared to have been set out to run sinuously between these intervisible points so as to provide the troops manning the Wall with the best advantage of the ground when facing aggression from the north.

To the author, this appears to have been a much more sensible way of planning a barrier to prevent unsolicited incursions into Roman territory than that which had been adopted for Hadrian's Wall. This was why, in retrospect, the author had found himself somewhat surprised to find that road-planning tactics appeared to have been employed in setting out the line of the latter. However, the important point here is that, along the Antonine Wall, the author was not able to find long-distance alignments, even though, by this time, he was looking out for them. This seems to strengthen the likelihood that the existence of such alignments is not imaginary. In other words, when long-distance alignments had been detected by the author, it is very likely that they had indeed been in existence in Roman times. Conversely, where they couldn't be detected, they probably had not existed.

In view of some doubts which have been expressed about the author's interpretation of the planning processes for Roman roads, it may be instructive to digress at this point to consider how and why Roman planning practices may have differed from those of more recent times. Standard practice in setting out the courses of new roads nowadays is described by Dr Hugh Davies in his book *Roads in Roman Britain*.[3] Dr Davies had worked in post-war road planning before obtaining his archaeological qualifications at Reading University.

Typically, the modern planning process would involve surveying all of the landscape to be traversed by the new road, and then working out three or so alternative lines across it in detail. This would be undertaken in order to assess the relative costs and difficulties of construction, as well as the comparable advantages of each option. These alternatives would then be presented for public consultation and for any political and other factors to be taken into account before a final choice of the preferred line would be taken. Essentially, all of the surveying would have been carried out in order to ensure the affordability of the construction, and to minimise the costs of construction and operation whilst maximising the value and benefits of the new road to its potential users.

Records show that much the same process of surveying and then comparison of alternative routes had been used by the surveyors and engineers who planned the courses of the commercial canals, railways, and roads that have been constructed in Britain over the last two to three centuries. In each case, there would have been sound engineering reasons for the courses that were eventually adopted. The effects of these are shown schematically in Figure 14, where the typical courses of each type of structure between two non-intervisible points, A and B, are indicated.

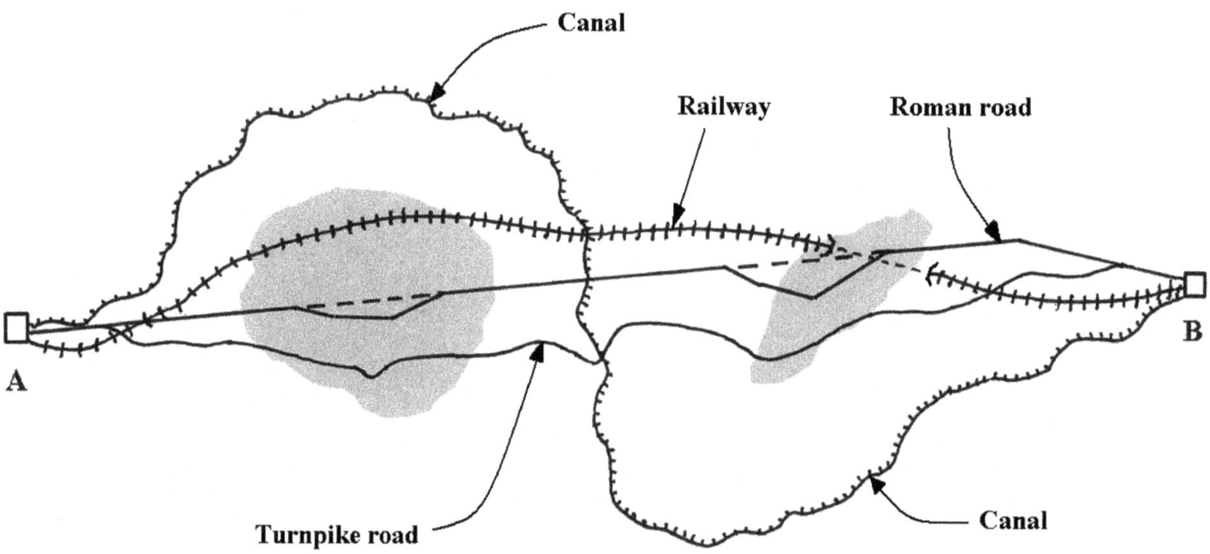

Figure 14 Diagram illustrating the different courses that might be taken by a canal, a railway, and a turnpike road between two non-intervisible places, A and B, according to their differing engineering requirements, and contrasting these with the course that might be taken by a Roman road across the same landscape

[3] Davies, 2002, 44

With canals, for instance, the water in them lies flat, so that they would have had to adopt a level course except where they could rise or fall by means of locks. Locks, however, were to be avoided if possible because they were expensive to construct and maintain, and hindered the flow of traffic. In addition, canals had to be designed so that their summit levels were as long as possible so as to provide an adequate reservoir of water. This is because each boat that left the summit level – in whichever direction – would take a lockful of water with it. Hence a heavy flow of traffic through an inadequately large or watered summit level could soon run it dry and close the canal. The outcome of the resultant need for long level stretches on canals was that they tended to follow circuitous courses around the contours of the countryside. This was especially so if extensive cuttings and embankments were to be avoided, which was the case with many of the earlier canals.

Railway lines, in contrast, did not need to be level, but, especially in their earlier days, the gradients on them had to be relatively gentle because of the limited power-to-weight ratios of the steam engines over the trains that they were pulling. In addition, tight curves had to be avoided if the trains were not to come off the rails at high speed. The result was that railways tended to be much straighter than canals, but they had to be constructed, in consequence, with much larger cuttings and embankments, coupled with many more tunnels and viaducts along the way.

With the turnpike (i.e. toll) roads that were constructed at approximately the same time as the canals and the early railways, the rolling resistance of even iron-shod wheels upon their gravel surfaces was much higher than that of such wheels upon iron rails. For instance, although a single horse might be able to pull a 60-ton barge upon a canal, or up to six laden trucks on a railway, it might take six or even more horses to pull a 2-ton wagon along a gravelled road.[4] In consequence the vehicles using such roads required a much higher power-to-weight ratio than on railways, and this meant that gradients on the roads would have been, in proportion, less of a handicap to travel. In addition, speeds would have been much lower than on railways because nearly all of the vehicular traffic was drawn by animals. The consequence of these factors was that turnpike roads could possess steeper gradients and sharper turns on them than railways, but this meant that their costs of construction could be proportionately lower.

Modern motor roads and motorways, on the other hand, have to be designed for the safe and steady flow of combinations of different types of fast-moving vehicle, from high-powered cars and motorbikes to heavily-laden lorries and trailers. For these kinds of traffic flows, the needs are for generally gentler gradients and curves than on turnpike roads. The gentler gradients are required so as not to disrupt the flow of traffic up hills between the various types of vehicle, whereas the gentler curves are required, not only to allow higher speeds, but also to ensure that drivers' sightlines are adequate, such that vehicles would be able to stop in time when spotting stationary traffic ahead.

[4] Gerhold, 2005, 52

Thus, in each case, the course adopted for each of these types of structures can generally be understood readily in terms of their engineering requirements. This is not the entire picture, of course. It must be remembered that in Britain, nearly all of the canals, railways, and turnpike roads were built as private ventures by investors who would want to obtain a return on their investments. Thus the courses of their constructions between the two end-points A and B could be influenced by any sources of potential traffic, such as mines, large industrial works, and settlements of any significant size, which they might seek to visit *en route*. In addition, objections from powerful landowners might force the most favourable route in terms of engineering to be avoided. Nevertheless, in between these points and around such obstacles, the economic case for achieving the most effective engineering for their particular structures would have loomed large, and even today, although modern motor roads and motorways tend to be funded from the public purse, the costs and ease of construction still remain paramount.

In contrast, with at least some Roman roads, there appears to have been little or no concern about the costs of construction or the ease of use of their roads by vehicular traffic. As is indicated on Figure 14, the lines of these roads tend to run straight across the countryside, often regardless of the topography, with possibly just a slight change of alignment where the final destination comes into view. Here and there a few relatively short deviations might be inserted to circumvent unavoidable obstacles or to reach local objectives, but essentially the intention appears to have been to keep the line between the main destinations as direct as possible. This design intention appears to have been primarily military – to enable marching and mounted troops to move across country as rapidly as possible. Moreover, the application of this intention hints at a world in which labour was both cheap and plentiful and where the scarce commodity, if there was one, might have lain in long-distance surveying skills and capabilities. Thus it is evident that the design criteria and demographic contexts in which roads were laid out in Roman times were likely to have been quite different from those of modern times.

In these circumstances it is, perhaps, surprising that people may have assumed that road planning in Roman times would have been similar to today's practice. Even from a glance at modern-day maps, it can be seen that the Romans were working to a different set of criteria from those of more recent times. In the way that their roads stride across the maps in very direct lines, they stand out as being quite different from those of any of the canals, railways or roads on the maps today – with one exception. That exception relates to those roads which were laid out when tracts of the countryside were enclosed in Britain in the 18th and 19th centuries. Like Roman roads, these enclosure roads often give the appearance of having been laid out by drawing a straight line with a ruler across the map. However, unlike Roman roads, enclosure roads tend to be relatively short in length and not to be underpinned by any long-distance alignments.

To return to the discoveries made by the author along Dere Street and elsewhere, the possibility is striking that – from the available evidence – there might in places have been a lapse of several years between the laying-out of the long-distance alignments and the eventual construction of what was clearly an important main road. Moreover, from detectable differences in the calibre of the surveying, it appears likely that different teams – or even generations – of surveyors may have been involved in the two stages of the process. Hence the possibility arises that these long-distance Roman alignments may not have been set out solely for the purpose of planning roads, but for other reasons as well.

For instance, the long-distance alignment that appears to underpin part of the course of Dere Street in the Vale of York was also found to have been used to fix a small portion of the course of the Rudgate Roman road, which clearly branches off the built line of Dere Street at Whixley and therefore was apparently planned later. This implies that the long-distance alignment was not created and used as a one-off exercise but was marked out and retained for future reference. See Figure 15.

In addition, it was observed that at least two Roman forts (Piercebridge and Ebchester), and possibly a third (Cappuck), were found to lie exactly on the long-distance alignments that had been projected as underpinning the course of Dere Street. Moreover, where they lay on the alignments was where the course of Dere Street, as built, had deviated somewhat from the alignments. This may perhaps suggest that the locations of the forts had been the primary focus for the Roman planners. See Figures 16, 17 and 18.

Hence the question was posed: might these long-distance alignments have been more than merely devices for setting out the courses of roads? In the BAR 492 monograph, the question was presented as a curiosity for readers to consider. By the time that the Amberley book came to be written, some months later, further reflection had made it clear that the question was not just a matter for curiosity: it was a question for serious research.[5] The sections of the monograph that follow set out to address this question and then describe the processes by which some of the alignments, at least, may have come to be used to guide the setting-out of certain Roman roads in northern Britain.

The conclusions, it should be noted, are confined to northern Britain because that is where the bulk of the author's research has been conducted. It may be considered very likely that the conclusions would have applied throughout the whole of Roman Britain, and quite possibly to elsewhere in the Roman Empire, but the author has not conducted research in these areas and so considers that it would be inappropriate to make such claims.

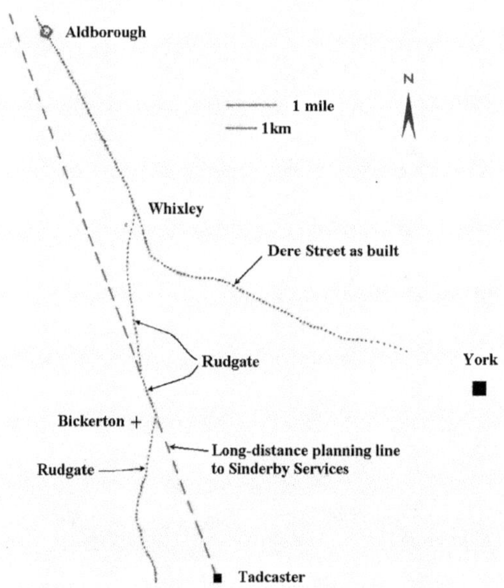

Figure 15 Part of a long-distance alignment for Dere Street being re-used by the Rudgate Roman road

[5] Poulter, 2010, 77

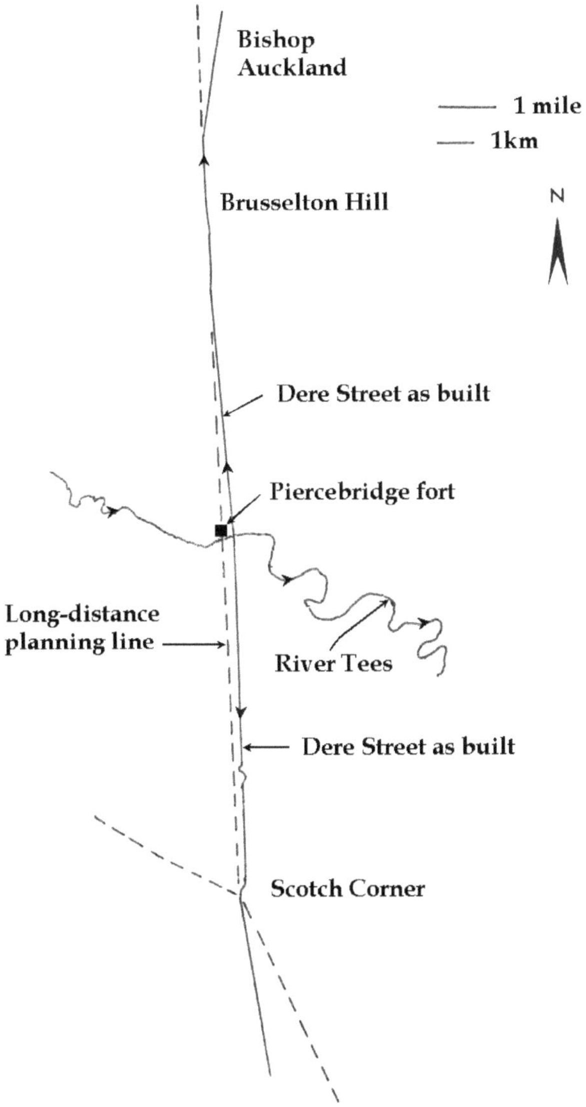

Figure 16 The Roman fort at Piercebridge standing on the long-distance planning line for Dere Street north of Scotch Corner

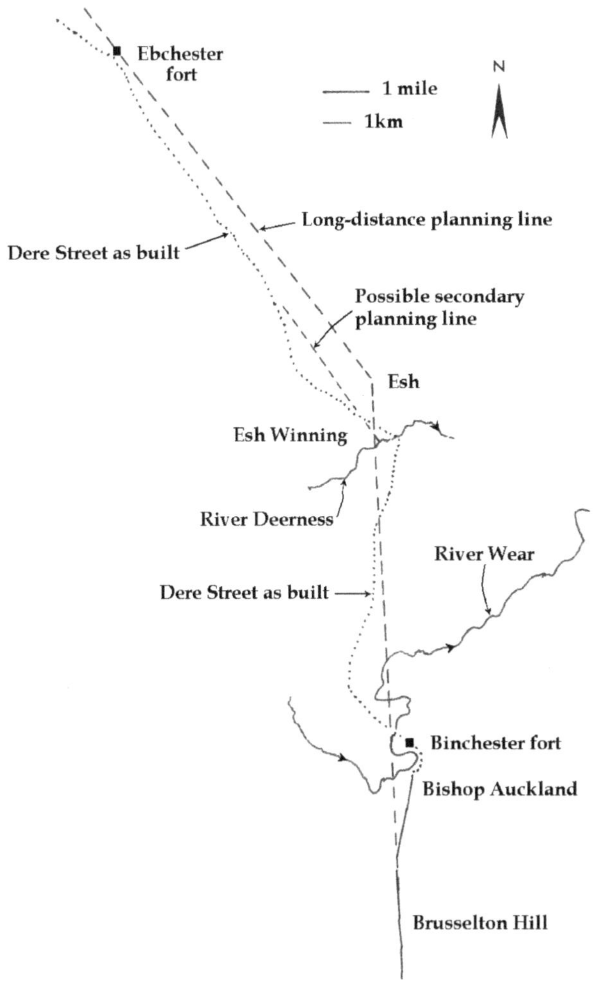

Figure 17 The Roman fort at Ebchester standing on the long-distance planning line for Dere Street that ran north-westwards from Esh towards the Dry Burn. See also Figure 11

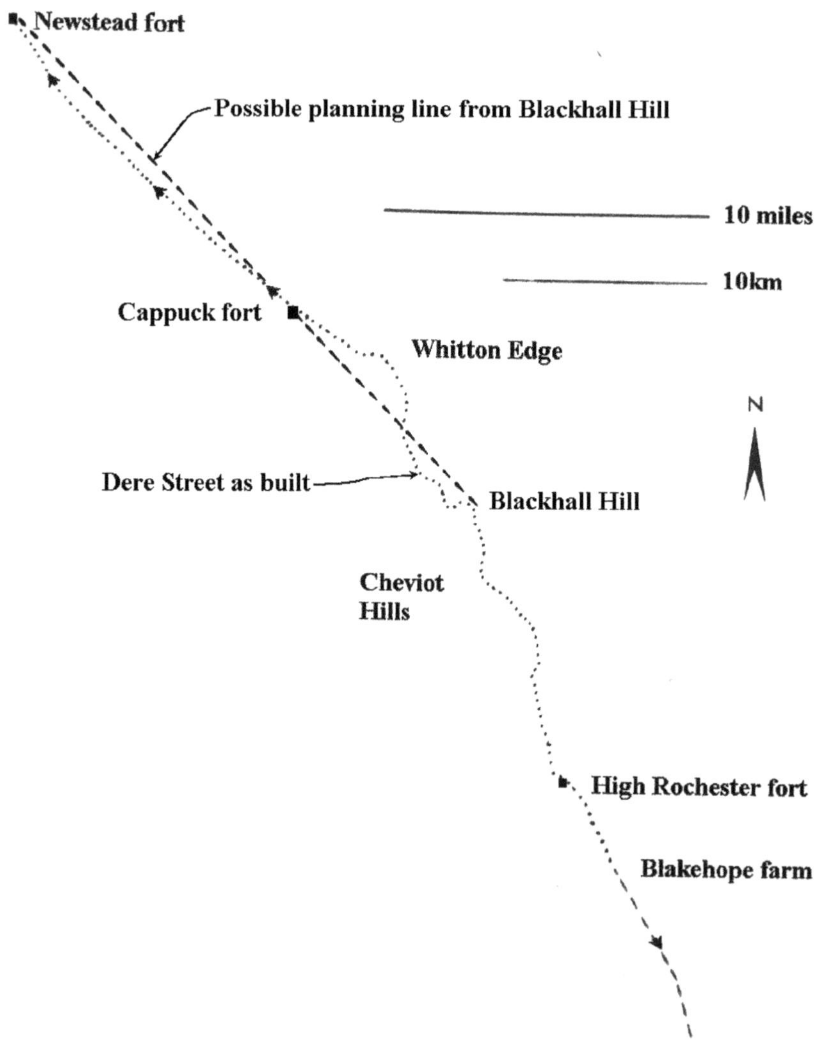

Figure 18 Position of Cappuck fort on a possible long-distance planning line for Dere Street between Blackhall Hill and the fort at Newstead

Section 2: New evidence received

Thus this was the question which had been left outstanding from the author's previous research: what might have been the purpose of the long-distance alignments, of the kind which the author had detected underlying the course of much of Roman Dere Street? When the author proposed that the question should be treated as a matter for further research rather than simply as a curiosity, he had no particular intention of attempting to address the question himself. The hope was that other scholars and students of Roman planning, perhaps with a wider knowledge of the Roman world, would pick up the question and examine what the purpose or purposes of these alignments might have been.

However, as it happens, just one day after the author's Amberley book had been published, David Breeze put the author in touch with an independent researcher called Rob Entwistle, who seemed to have discovered a whole network of long-distance Roman alignments running across northern England. Rob Entwistle had made these discoveries quite independently of the author, and by an entirely separate means: he had been investigating aerial images provided by Google Earth. When examining the landscape of northern England via Google Earth, Rob Entwistle had been surprised to find out how many Roman roads and sites appeared to line up over very long distances. Moreover, he had observed that many of these alignments seemed (a) to fall into clusters, and (b) to run at set angles to each other.

Despite having identified significantly more potential alignments than the author, Rob Entwistle had not recognised the ones that the author had discerned underpinning Dere Street. Therefore the first question to be addressed was whether what Rob Entwistle and the author had found were examples of the same phenomenon or not. After some days spent examining each other's findings, it was mutually agreed that they did indeed appear to be instances of the same thing, albeit that they had been detected by entirely separate processes of enquiry.

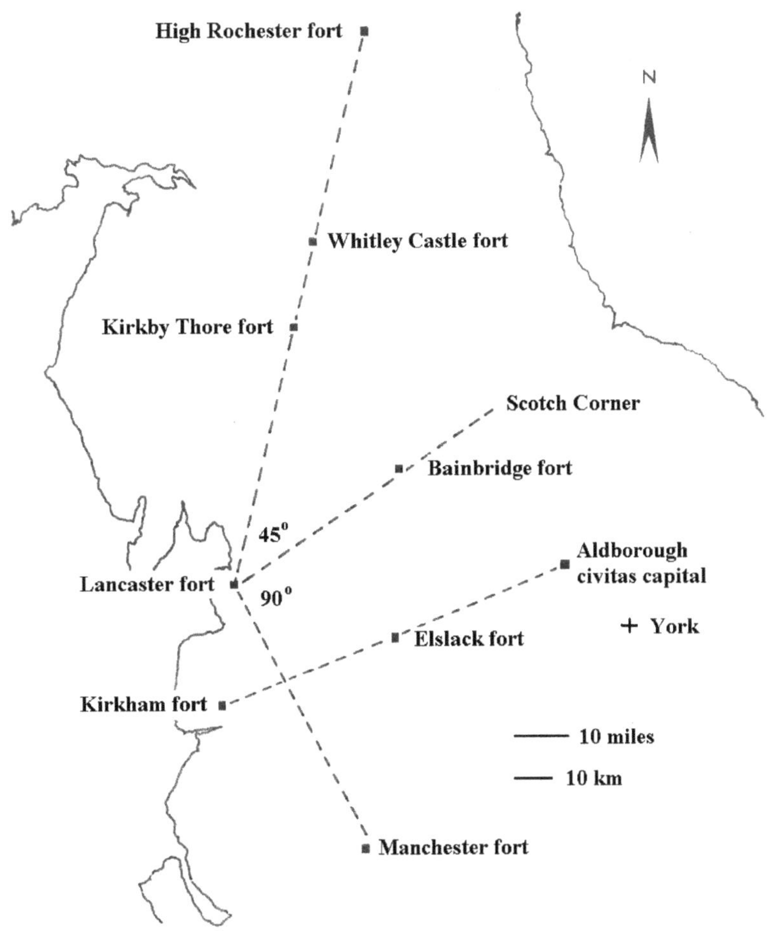

Figure 19 A small selection of the long-distance alignments detected by Rob Entwistle

Figure 19 shows a small subset of the long-distance alignments that had been detected by Rob Entwistle. Later in this monograph, in Sections 7 and 8, these particular alignments will be used to illustrate how the author believes that two of the three roads which he has selected as examples were set out, using parts of these alignments in the process. It is to be hoped that Rob Entwistle will soon find a vehicle to publish his findings in full, so that scholars can examine the overall picture to be revealed by his discoveries.

In the meantime, the second question to be addressed was whether or not such a network of alignments might be imaginary. Archaeologists in Britain will be familiar with the concept of ley lines, and the notion that people in prehistoric times laid out dead straight lines across the landscape in order to guide travellers on their long-distance journeys. As was discussed in the author's Amberley book, this concept was shown to be spurious by Professor Tom Williamson and Liz Bellamy, being based upon flawed analysis and unfounded assumptions.

In the process, however, Tom Williamson and Liz Bellamy made the important point that, statistically, it is not difficult to draw a long straight line across the landscape of Britain and to have it touch or run close to sites of any sort that one might consider meaningful.[6] In the context of a network of Roman alignments, this would include Roman forts and settlements.

Expanding this point from the linear into two or more dimensions and putting it more generally, Matt Parker, a mathematician at Queen Mary College, London, has declared that any number of patterns can be discerned to any specified degree of accuracy in any sufficiently large set of data.[7] He illustrated this point by discerning various nonsensical alignments and geometrical relationships in the distribution of Woolworths' stores in England and Wales. See Figure 20. With regard to the context of this monograph, it cannot be denied that there were many more Roman sites in Britain than there were Woolworths' stores.

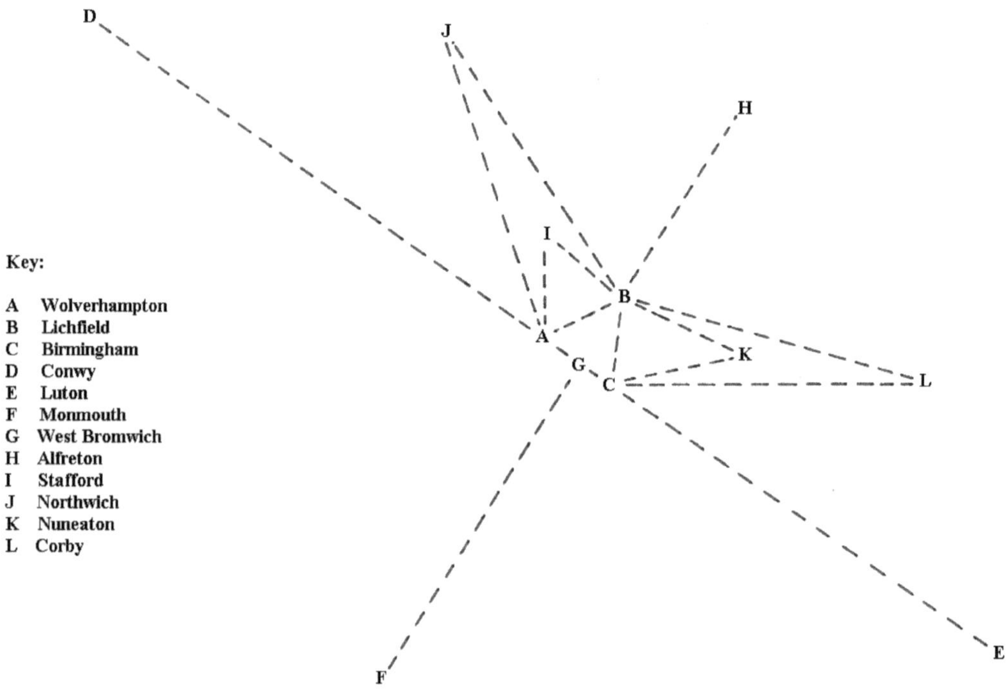

Figure 20 Patterns detected in the locations of Woolworths' stores by Matt Parker. To quote from his Press Release: "Three stores around Birmingham (Wolverhampton, Lichfield and Birmingham) formed an exact equilateral triangle, and if the base of the triangle is extended it forms a 173.8 mile line linking the Conwy and Luton stores. Despite the distance involved, the Conwy Woolworths' store is only 40 feet off the exact line, and the Luton site is within 30 feet. All four stores align with an accuracy of 0.05%. The bisector of this same triangle then passes through the Monmouth, West Bromwich and Alfreton store locations with an accuracy of 0.5%. There are also grids of isosceles triangles on each side of the Birmingham Woolworths' triangle. One such isosceles triangle made with Stafford only has an error of 3% and it points directly at the Northwich Woolworths' store that is itself only 0.6% off being exactly isosceles. These patterns were found from the 800 random locations of Woolworths' stores by simply skipping over the vast majority of the sites and only choosing the few that happen to line up."

[6] Williamson and Bellamy, 1983, 94-5 and 97-102
[7] Parker, 2010

Faced by such fearsome admonition, and with the unfortunate precedent of ley lines well in mind, it must seem most unwise to try to argue that the Romans had set out a series of long straight lines across the countryside. Yet, as discussed in Section 1, from extensive analyses of the course of Dere Street, the author had found that the planning of some parts of that Roman road could not be explained without the existence of underlying long-distance alignments. Moreover, from the ways in which certain Roman roads, as built, relate to some of the alignments discovered by Rob Entwistle, the author is convinced that those particular alignments, at least, did exist in Roman times and were used in the manner to be described and illustrated later in this monograph.

Thus, from the author's point of view, although it is always possible that some of the alignments discovered by Rob Entwistle could be coincidences, certain of them are considered – with some confidence – not to be so, and these, taken together, add to the likelihood that one or more networks of long-distance alignments did exist in Roman times. To conclude on this note, the point should be repeated that neither the author nor Rob Entwistle were looking for such patterns of alignments when they first discovered them. In both cases, these discoveries were unexpected and occasioned surprise.

Given the possibility, at least, that one or more networks of such long-distance alignments may have existed in Roman times, this still leaves the question of what the purpose or purposes may have been for their creation. Initial thoughts that they may have been the vestiges of triangulated surveying exercises were scotched by Dr M. J. T. Lewis,[8] with whom Rob Entwistle had been in correspondence prior to his contact with the author. As Michael Lewis had pointed out, Roman surveyors did not possess the optical instruments that would have been capable of registering angles in degrees, and these would have been necessary for large-scale triangulation exercises. Moreover, even with the superior equipment that surveyors possessed in the 19th century, it had taken the Ordnance Survey some 70 years to cover the whole of the country, whereas it can be imagined that the Roman military authorities would have desired results more quickly than that.

After some deliberation, the likeliest conclusion that was reached was that at least some of the long-distance alignments may have been the relics of survey lines that had been set out in order to form rapid assessments of the topography and demography of large areas of countryside. This would have been in connection with the acquisition of new territory by the Roman army. This conclusion was developed in correspondence between the author and Rob Entwistle and only a brief summary of its substance will be offered here. The full argument and evidence for this conclusion will be presented in Section 5 of this monograph, along with a discussion about the implications for Roman mapping and about the military and political information which would have been available to Roman governors and generals.

In the meantime, and in a nutshell, the model that was conceived is that, in connection with the acquisition of new territory, the Roman military authorities would have needed rapid and reliable assessments of the countryside – not just of its topography (its hills, valleys, major rivers, woodland, marshes, trackways, etc.) but also of its demography (its native population, hostility, languages, groupings and relations with neighbours, settlements, agriculture, etc.). To obtain these assessments, it was suggested that, at times, survey parties would be sent out on a number of bearings at set angles to each other. As indicated in Figure 19, such set angles might, for instance, be right angles or angles of 45°, which could be easily constructed using basic geometry both in the field and at a desk. Such set angles would thus be used so that the reports from the survey parties could be recorded with meaningful geometrical relationships to each other back at the military authorities' headquarters.

These reports could then be used, in the period of consolidation after the conquest, to determine where the network of forts and roads needed to be established in order to maintain the Romans' grip on the newly-conquered territory. These decisions would then need to be relayed to the commanders on the ground, and the simplest and surest way to do this would be to do so with reference to the survey lines which had been established. This would not necessarily mean that the forts and roads would be obliged to be sited on the survey lines, of course. However, in some cases it might have been found convenient to locate some forts on or close by the lines that the surveyors had been following. In such cases it would have been easy for the road planners to know where the roads should run between them.

It is appreciated that this model may not represent the complete picture. The author's concern lies primarily in trying to understand how Roman roads were set out, and it is possible that the long-distance alignments may have had other functions too. For instance, such survey lines might have been used to demarcate administrative boundaries or areas of responsibility. As proposed in Section 5, they may also have been set out to be used as guidelines for the supply trains and lines of communication that would have followed in the wake of the advancing armies. No doubt Rob Entwistle will be giving consideration to these questions when the full corpus of his findings becomes available, which is why its publication is so much awaited.

[8] M. J. T. Lewis, *pers. comm.* to Rob Entwistle, 6 May 2010.

Section 3: Tools used for the observation and analysis

Before proceeding to examine the evidence and examples in detail, it may be appropriate to inspect some of the modern-day tools which have been used for the observation and analysis of much of the material in this monograph.

For instance, Roman roads are famed for their straightness, even though in fact not all of them are straight. Nevertheless, the ability of Roman surveyors to set out very straight long-distance alignments, when they chose to do so, is well known. In Britain, the first 12½ miles (20 km) of the Roman road from Roman London Bridge to Chichester is aimed exactly at the east gate of Chichester, nearly 60 miles (96km) distant.[9] In Germany, the Outer German *Limes*, between Walldürn and Haghof, runs almost dead straight for more than 50 miles (81km), apart from a one-mile (1.6km) deviation to avoid a steep valley.[10] So why, it might be asked, have so few of these very long-distance alignments been discovered up to now?

At least part of the answer, it is considered, lies with the advent of digital technology. Via facilities such as Google Earth and digital mapping, it is possible to zoom from inspecting minute details on the ground to viewing entire landscapes more than 100 miles (160 km) across. This makes it much easier to discern alignments and relationships which are more than, say, 30 miles (48 km) apart. In contrast, with a 1:50,000 printed Ordnance Survey map, for example, the field of observation is limited to about 25 miles (40 km) square. Although this limitation does not wholly prevent the detection of longer-distance alignments, it does make it much more difficult to spot or notice such extended relationships by means of casual observation.

In addition, the digital technology allows straight lines to be drawn across the images. This allows alignments to be checked and the intervisibilities of points to be tested almost instantaneously. This is no substitute for visiting and checking alignments on the ground as well, but it makes it much quicker and easier to compare and eliminate possibilities before pursuing particular lines of enquiry. Thus digital technology gives the researcher much greater powers to examine wider landscapes both more rapidly and with greater scope than had been available to him or her in the past. It is this, it is believed, which is one reason why a researcher such as Rob Entwistle has been able to detect alignments which may not have been recognised in the past.

[9] Lewis, 2001, 238-42

[10] Lewis, 2001, 242-5

Another new facility which is becoming increasingly available via digital technology is LiDAR. This technology enables tree cover and other obstructions to be stripped away digitally from aerial images so that ground surfaces can be examined for indications of former structures such as the foundations of buildings and the courses of roads. The technology can also differentiate between the heights of crops, etc., so enabling features below the surface to be investigated as well. This functions in much the same way that crop marks can be used, but without having to wait for favourably dry weather. Hugh Toller's recent work on the course of the Western Main Roman Road north of the Lune Gorge illustrates what a powerful tool LiDAR can be and highlights its potential for further research into the courses of Roman roads which are currently considered to be lost. See Section 7 for the new course of the Western Main Roman Road which has been revealed by LiDAR.

Once the available evidence has been gathered, by whatever means, the author then makes use of a systems-based modelling methodology in order to analyse the archaeological evidence, and interpret what may have been happening in previous times. This modelling tool has been imported from management science, and examples of its use in archaeology have already been described and illustrated at length in the author's Amberley book, as a result of which the full detail will not be repeated here.[11]

However, to characterise the methodology briefly: rather than just considering different ideas as options about what may have happened in the past, the full systems implications of each idea are worked out by modelling. Then, not just the ideas but the systems models are compared with the available evidence to determine which might be the likeliest option.

> Thus, rather than proposing that, say, the Romans might have done *a* or *b* or *c*, the analysis is conducted by modelling the systems consequences of such actions.
>
> For example, if the Romans had been doing *a*, then it can be modelled that they would also need to have done *l*, *m* and *n* in order to make *a* work.
>
> Alternatively, if the Romans had been doing *b*, then it might be modelled that they are very likely to have needed to do *p* and *q* as well.
>
> On the other hand, if the Romans had been doing *c*, then it might be modelled that they would have

[11] Poulter, 2010, 111-16

needed to do or are very likely to have carried out *x*, *y* and *z* in order to have made things work.

Thus, instead of simply comparing *a*, *b* and *c* with the available evidence, the occurrences of

- *a* with *l*, *m* and *n*,
- and *b* with *p* and *q*,
- and *c* with *x*, *y* and *z*

would be compared with the evidence.

Applying the methodology in this way enables much more powerful examinations to be conducted of what may have been happening in the past. Sometimes archaeologists undertake such modelling naturally or unconsciously, without recourse to a formal methodology, but perhaps the difference is that the author attempts to do it consistently. The detailed arguments for the existence and use of survey lines in Sections 5 and 6 are examples of the use of fully worked-out systems models.

Nonetheless, in order for such systems models to be realistic, the persons who create them have to be good judges of how people would behave when faced with certain situations. This is a skill which is something that is not, as far as the author is aware, taught in schools or universities. It may even be a gift. Either way, it is an essential capability for any person who is going to design a system which is going to work well in practice, whether it be a transport system, a financial system, an educational system, or a new computer system – all of which will, of course, have human interfaces.

Such a capability is not a matter of intelligence. The difficulty in applying it to the interpretation of archaeological evidence, however, is that it can be hard to judge its efficacy in this field.

With regard to the archaeological interpretations developed in this monograph, the author considers that he does possess such skills and has a track record to support this belief. The track record includes nearly 40 years' professional experience of designing new systems and ways of working in a variety of environments, both in Britain and abroad, virtually all of which were successful right from their initial implementations. In the field of archaeology, as described in Section 1, it was the author who proposed how the planners of Roman roads might behave when seeking to make a change of alignment, and when tested against the evidence on the ground, this supposition was confirmed as very likely to be correct.

In drawing out this point, it is not intended in any way to seek to be boastful. Different people bring different skills to the field of archaeology. The reason for reporting it here is simply to register, for the sake of completeness, that it is part of the toolkit which has been employed to interpret the evidence which is described in this monograph.

Section 4: Tools used by the Roman surveyors and planners of roads

Before delving into various possible reasons for the creation of their long-distance alignments, it is worth reviewing briefly some of the instruments which were available to the Roman surveyors and planners of roads.

The best-known surveying instrument which may have been used by the Romans for planning the courses of their roads is the *groma*, although the exact details of its operation remain unclear.[12] Figure 21 shows a reconstruction by Matteo Della Corte of an example of such an instrument which had seemingly been found in pieces during excavations in Pompeii.[13] It was discovered there amongst the ruins of one or more houses, along with a set of other instruments that also appeared to have been used for surveying. More recently, the details of this reconstruction of the *groma* have been challenged,[14] and an informed discussion of the various issues concerning this disagreement is offered by Michael Lewis in his *Surveying Instruments of Greece and Rome.*[15] Notwithstanding these uncertainties the key point for the purposes of this monograph is that, by sighting between the plumb lines of the instrument, it can be seen that it could most readily be used for extending straight alignments and for setting out lines at right angles.

For setting out lines at right angles, John Peterson has pointed out that the cross-piece of a *groma* could not have been made with sufficient accuracy to be able to set out a true right angle. A true right angle might have been achieved, he proposes, by rotating the *groma* through a quarter turn after taking one sighting, and then taking a second sighting and then splitting the difference between the two sightings.[16] This would have compensated for any error in the cross-piece of the instrument.

In September 2011 the author was in discussion with Mr David Robson, a surveyor and engineer with practical experience of setting out new roads in northern Nigeria. Mr Robson had subsequently developed a keen interest in the subject of Roman roads, and the author is indebted to him for the proposal that the Romans carried out much of their long-distance surveying by night. Mr Robson drew the author's attention to the presence of a lamp amongst the set of what appeared to have been surveying instruments which had been found at Pompeii,[17] and remarked that from his own experience it was astonishing how far away a light could be seen at night. Since that meeting the author has been given to understand that the early Ordnance Surveys in Britain were also conducted using lamps at night, and this is confirmed in General Roy's account of his pioneering work on landscape surveying in Britain in the late eighteenth century.[18] In addition, in a comprehensive review of possible Roman surveying practices, Isaac Moreno Gallo similarly concludes that the Romans are likely to have used lights at night when surveying over very long distances.[19]

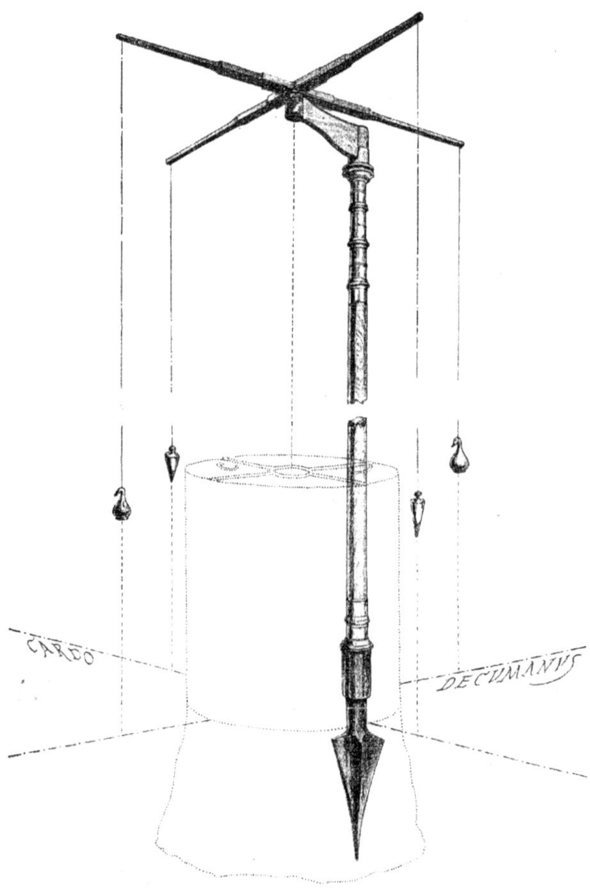

Figure 21 Reconstruction of a groma found at Pompeii

It will be noticed that on the reconstruction of the *groma*, the plumb weights have two different shapes. It may be, therefore, that if the Romans had indeed been undertaking long-distance surveys at night, then the different shapes

[12] Lewis, 2001, 124
[13] Della Corte, 1922
[14] Schiöler, 1994
[15] Lewis, 2001, 128-33
[16] Peterson, 1993, 58-9
[17] In his booklet describing what is to be seen from the excavations at Pompeii, Matteo Della Corte mentions "a large and curious phallic lamp" which was found in the House of Verus along with the remains of the *groma*; see Della Corte, 1925, 17.

[18] Roy, 1790, on several pages, but especially 265-6. I am indebted to Humphrey Welfare for this reference.
[19] Moreno Gallo, 2004, pages 41-2 in the English translation by Bishop, 2006

at the ends of the plumb lines would have enabled the surveyors to ensure, by feel in the dark, that they had not accidentally rotated the instrument through two quarters rather than one. It must be remembered, though, that the reconstruction shows only one way in which the plumb weights may have been arranged. If they had been arranged so as to be adjacent to each other rather than diagonally opposite, then this suggested explanation might not make quite such a good fit.

However, there are some uncertainties about the extent to which the *groma* would have been used by the Romans for their long-distance surveying work. On the grounds of the instrument's perceived lack of precision, Isaac Moreno Gallo goes so far as to dismiss its use altogether for such purposes.[20] For setting out long-distance alignments, he appears to favour the possibility that an alternative surveying instrument known as the *dioptra* would have been employed,[21] and Michael Lewis concurs that the slit sights of a *dioptra* would not have been prone to disturbance in windy conditions, unlike the dangling cords of the *groma*.[22]

This brief overview, however, is not the place to enter the debate about what kinds of surveying instruments were known to the Romans, and about how they may have been used. For full accounts of the *groma,* the *dioptra* and other Roman surveying instruments, the reader should turn to the publications by Lewis and Moreno Gallo.[23] For the purposes of this monograph, the key point is that, notwithstanding any uncertainties which there might be about what surveying instrument was employed, and in what way, there is abundant evidence on the ground that Roman surveyors were able to set out precisely straight alignments for very long distances over extensive areas of undulating countryside, and it is to the possible purposes of such alignments that this investigation will now turn.

[20] Moreno Gallo, 2004, pages 15-7 in the English translation by Bishop, 2006

[21] Moreno Gallo, 2004, page 44 in the English translation by Bishop, 2006

[22] Lewis, 2001, 132-3

[23] Lewis, 2001; Moreno Gallo, 2004

Section 5: Possible reasons for creating long-distance alignments

There are a number of possible reasons which can be postulated for the creation of such long-distance alignments. In order to appreciate the various circumstances in which they may have been created, it will be helpful to consider a model of the conquest of a new territory by the Roman army, and, in particular, the dispositions of Roman forts before, during and after that conquest period.

Within such a model, and with regard to the dispositions of the forts at different periods, the general sequence of events can be envisaged as follows:

1. the first forts would have been campaign forts, it is judged, sited in order to retain occupation of newly-conquered territory or to provide a springboard for its conquest in the next season. The locations of such forts would have been determined by the course of the campaign, i.e. by where the hostile forces may have been anticipated to be mustered. In general the lives of such forts can be expected to have been relatively short, and, as a result, their sites can be expected to have retained little or no upstanding remains. Thus the discovery of such forts is most likely to have come from aerial photography. Examples from the north of England might include the sites at Rossington, Long Sandall, and Roall, all of them south of the Humberside levels

2. the second forts, it may be considered, would have been established to protect the supply and communications routes as the campaigns advanced further ahead. The locations of these forts would have been at vulnerable positions along such supply and communications routes, especially at the crossings of major rivers. Examples in the north of England might include the sites at Roecliffe by the River Ure, Binchester by the River Wear, and Red House by the River Tyne, and possibilities can be suspected at Catterick by the River Swale and Piercebridge by the River Tees. Again, the lives of such forts can be expected to have been quite short except where the subsequent network of permanent all-weather Roman roads, when constructed, happened to make use of portions of the same routes. In these cases the forts along those portions would have had a much longer existence. Examples of such longer-lived forts could include Catterick, Piercebridge, and Binchester

3. it is envisaged that the third forts, and the network of permanent roads linking them, would have been established in the period of consolidation after the conquest had been completed, and would have been sited to as to maintain the long-term occupation and administration of the territory under Roman control. The locations of these forts and roads, it is proposed, would have been determined by surveys of the newly-conquered territories, taking account of (a) the distances between suitable sites and (b) the manpower available. In northern England, after the Roman invasion and then possibly, subsequently, in preparation for the withdrawal from Scotland, this period of consolidation could be seen as having been initiated in the early to mid 80s AD, and to have been marked, for instance, by the establishment of Aldborough and Corbridge in place of the forts at Roecliffe and Red House.

A first possible reason for creating long-distance alignments

The first possible reason for creating long-distance alignments relates to the second phase of the foregoing scenario. In some circumstances, as their campaigns moved forward, it may have been possible for the Roman army's needs for supplies and communications to have been met by means of existing roads and tracks, coupled with local knowledge of the countryside. Caesar, for instance, in his conquest of Gaul, appears to have been able to arrange for supplies to be moved into position without the need for any formal survey of the surrounding landscape.[24] In contrast, in Britain, the Roman army appears likely to have advanced quite quickly from northern England into north-eastern Scotland, and this may have resulted in unusually extended supply lines up the whole of northern England and across southern Scotland. It is considered that accurate local knowledge of such an extended stretch of territory would only have been partial at best. Hence setting out a series of long-distance alignments for the supply and communications lines to follow might, in this case, have been judged worthwhile.

Thus, in northern England and southern Scotland, one possible reason for the presence of such long-distance alignments may have been to provide guidance for the supply and communications lines that would need to have been in operation to support the army campaigning further north. No doubt, where practical, some of these needs for support would have been met by sea, and also along rivers, but land-based supplies and communications would have been more dependable, and therefore essential.

The model suggested is thus as follows. Once an army had conquered part of a territory and had moved forward to conquer more of the country ahead it would need to be sustained overland by supplies and support from the rear,

[24] David Breeze, *pers. comm.*

as well as to be in constant communication with military headquarters, also to the rear. Many of the persons operating such supply trains and communications lines overland would be travelling extensive distances across, to them, unknown landscapes, and in the absence of guidance there would be quite a risk of their getting lost, which could be both wasteful and hazardous.

Whilst existing trackways might be used where capable of bearing the traffic and heading in the right direction, their courses would be unlikely to suit the Roman requirements on all occasions. Yet at the time there would not have been the resources available to the Romans for them to construct their permanent all-weather roads. Those resources – the soldiers – would have been engaged on the campaign. Hence the trails that the supply and communications lines should follow in the interim would need to be planned and marked out on the ground as temporary routeways.

Thus an initial reason for creating such a framework of long-distance alignments may have been to set out the general direction which such temporary routeways should take and to guide where structures along their courses needed to be set in place. These alignments would have helped the Roman sappers, for instance, to decide where to build the river crossings, harden the ground underfoot or adapt existing trackways, and construct secure bases for overnight stays.

For the sappers, therefore, the key elements in the creation of such campaign supply trails would have been:

- selecting the crossing-places of rivers, which would, if possible, have been bridged by temporary structures and invariably guarded by nearby forts

- building hardened causeways or corduroys over soft, marshy, or churned-up ground

- constructing secure bases for stopovers at night.

It should be noted that a framework of such long-distance alignments could have been set out without requiring any extensive or formal surveys of the intervening ground. Scouting would have been sufficient to establish the lie of the land and identify the general direction to be taken. Once such alignments had been set out, however, then marking them out fairly closely on the ground would have provided guidance to where the various crossing-points, firm ground, and overnight stations were to be found for those persons travelling along the supply and communications lines, once they had been established.

The evidence for the existence of such a model can now be examined.

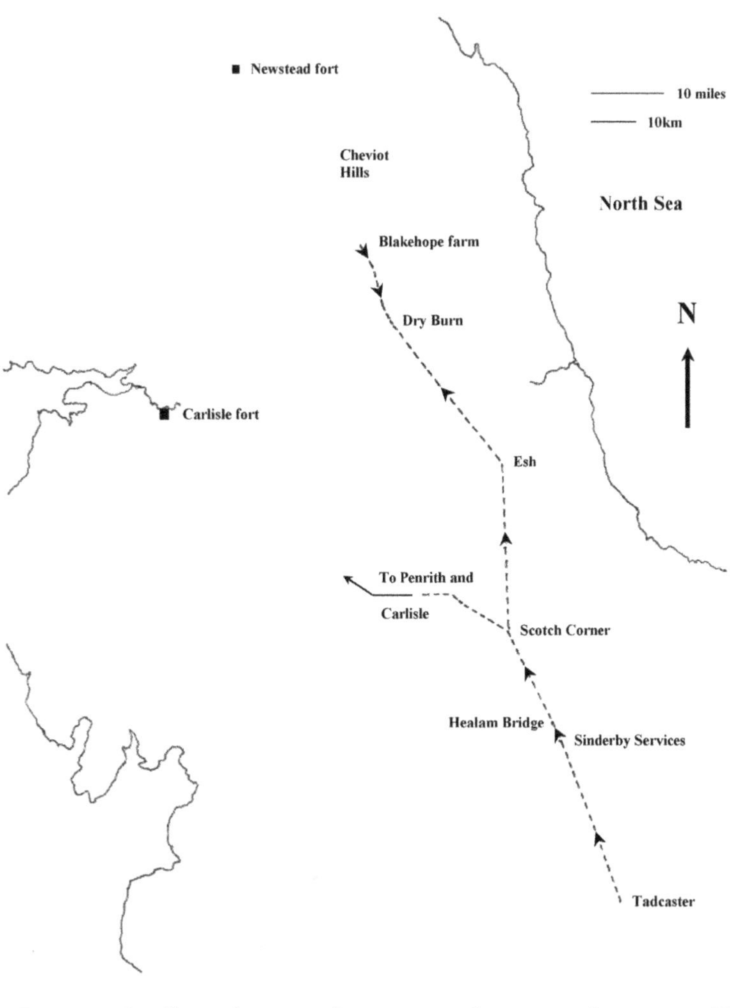

Figure 22 Framework of long-distance alignments underpinning the course of Dere Street

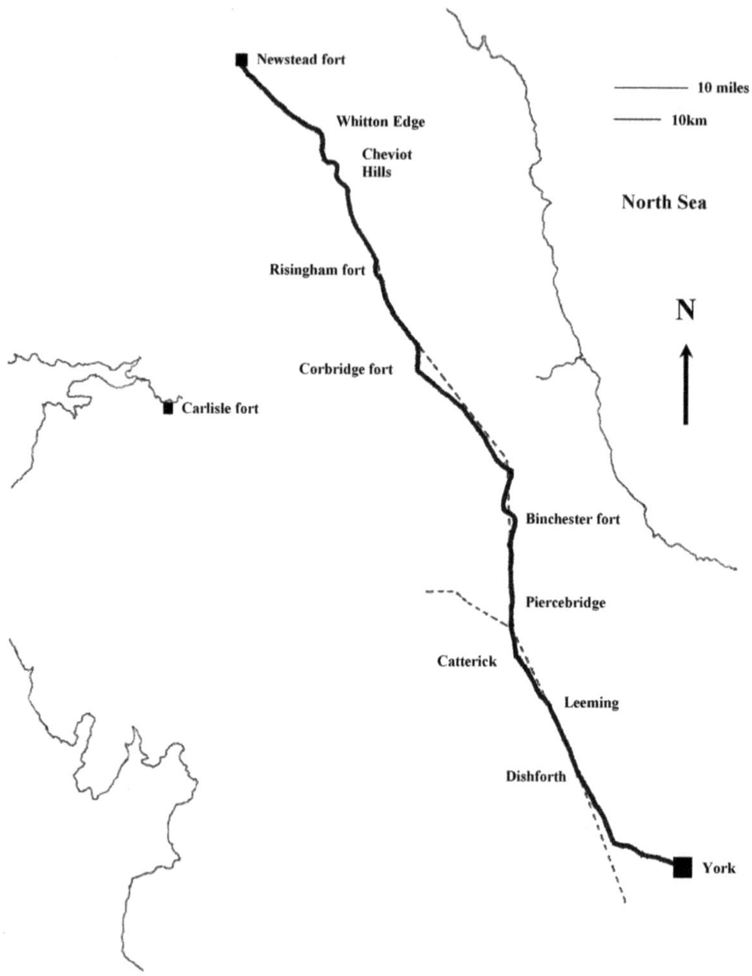

Figure 23 Dere Street as built, in relation to the alignments

Figure 22 shows the framework of long-distance alignments which were found to underlie the course of Roman Dere Street. The framework runs for more than 100 miles (160 km), from Tadcaster to Blakehope farm near the Cheviot Hills, and Figure 23 illustrates the course of Dere Street as built in relationship to this framework. It should be noted how little of the 24½ miles (39.5 km) long-distance alignment from Tadcaster to Sinderby Services was used when the permanent course of Dere Street was eventually laid out to run from York to the crossing of the Ure by Aldborough. There is thus already an indication that this southernmost long-distance alignment may have been laid out for some other purpose before the final course of Dere Street had been determined.

In fact, it is possible that the original southernmost long-distance alignment had started not from Tadcaster but from the vicinity of the aforementioned Roman site at Roall, some 12½ miles (20 km) further south.[25] Quite apart from the alignment, there is some evidence to support this contention. It depends upon the possibility that, starting from somewhere near Roall, there had been an original Roman supply route which had crossed the River Wharfe at Tadcaster and then followed the same long-distance alignment northwards to cross the River Ure near Boroughbridge. If so, then it is possible that the Rudgate Roman road is a relic of such a supply route, but that when the time came to construct the Rudgate as a permanent road, the choice was made to cross the River Wharfe upstream, to the west of Tadcaster – possibly because of persistent flooding – and then to pick up the original long-distance alignment near the village of Bickerton on its way towards Boroughbridge. See Figure 15. However, by the time that the Rudgate came to be constructed as a permanent road, it also appears likely that, as indicated in Section 1, the all-weather course of Dere Street had already been set out to run from York to cross the River Ure at Aldborough. Hence, rather than continue along the original alignment towards Boroughbridge, it appears that the course of the Rudgate may then have been diverted away from the old alignment, a mile or so north of Bickerton, so as to join the new line of Dere Street at Whixley. The relationships between the roads and their alignments would appear to fit this model quite well.

[25] I am indebted to Mike Haken for this suggestion.

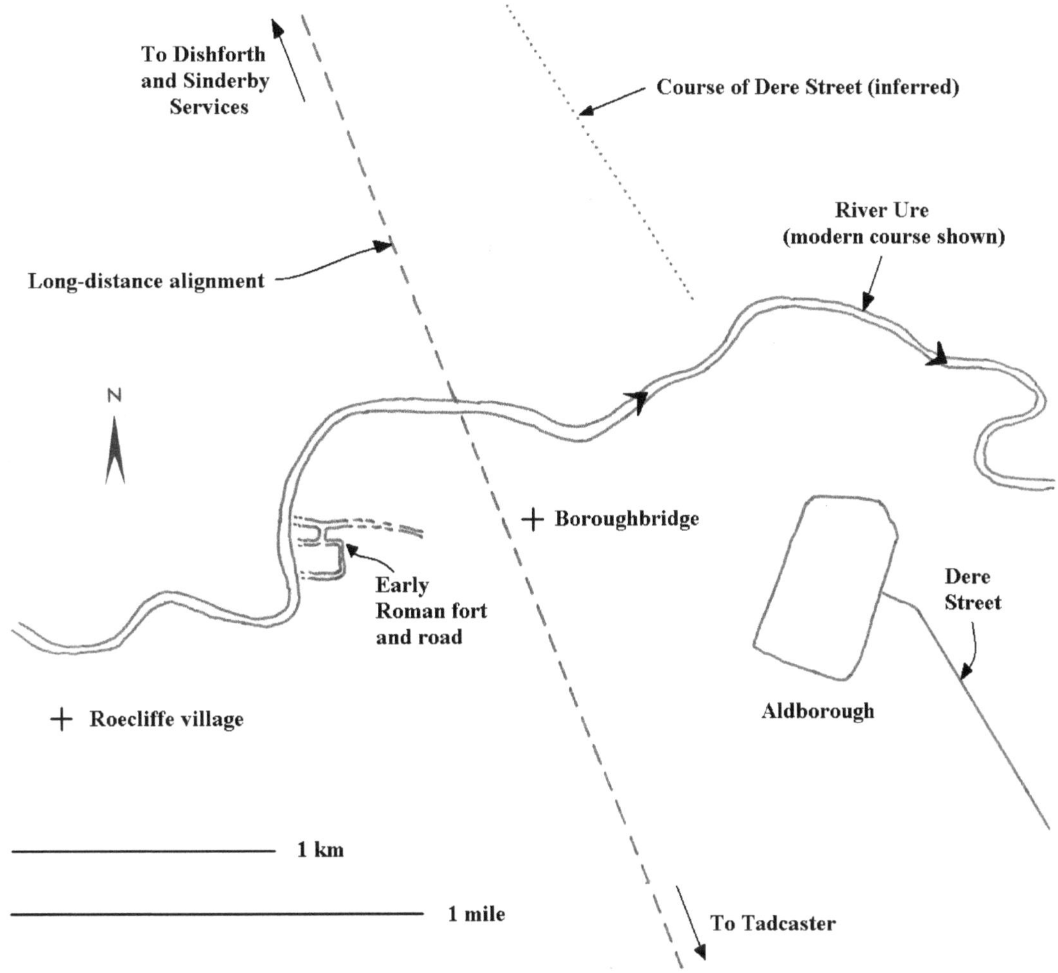

Figure 24 Location of the early Roman fort near Roecliffe in relation to the long-distance alignment from Tadcaster and the later course of Dere Street as built through Aldborough

Some 10 miles (16 km) further north of where the Rudgate branches towards Whixley from the long-distance alignment, Dr Mike Bishop has excavated what proved to have been a short-lived Roman military base sited beside the River Ure between Roecliffe and Boroughbridge, at a location which he has suggested might have marked an early crossing-point over that river.[26] He also raised the possibility that this crossing-point might have been part of a predecessor to the permanent course of Dere Street.[27] As can be seen on Figure 24, this point lies about ½ mile (0.8 km) to the west of where the long-distance alignment from Tadcaster would have crossed the Ure, but this digression may have been dictated by the propensity of the River Ure to flooding, not least in the Boroughbridge area. As it happens, the crossing-point at Roecliffe lies almost exactly upon a direct line from Tadcaster to Scotch Corner, but this should probably be counted as a coincidence because no other evidence for the use of such an alignment can be detected.

What Mike Bishop found appears to have been a Roman fort with an east-west road running just to the north of it and apparently down to a crossing of the River Ure. The nature of the crossing-point was not examined by excavation, but the road had been successively resurfaced and subject to at least three phases of development in a life of less than 20 years.[28] This suggests the convergence of heavy traffic upon a temporary crossing-point, which the fort had been positioned to protect.

Some 10 to 15 years after the fort near Roecliffe had been established, it appears that the crossing-point of the Ure had been relocated some 1¼ miles (2 km) or more downstream to near Aldborough. Presumably the occasion for this would have been when the permanent course of Dere Street had been laid out to run from York rather than Tadcaster, and aligned so as to cross the river at Aldborough rather than by Roecliffe, and then run on to join the original alignment near Dishforth. It was at this time that the original fort near Roecliffe (and probably the crossing-point, too) seems to have been abandoned.[29]

[26] Bishop, 2005, 217-18

[27] Although Mike Bishop also pointed out that it could have formed part of an early Roman route up Wensleydale, and noted that the possible existence of another early crossing-point of the Ure under the new course of the A1 road could not be ruled out.

[28] Bishop, 2005, 150 and 218

[29] Bishop, 2005, 218

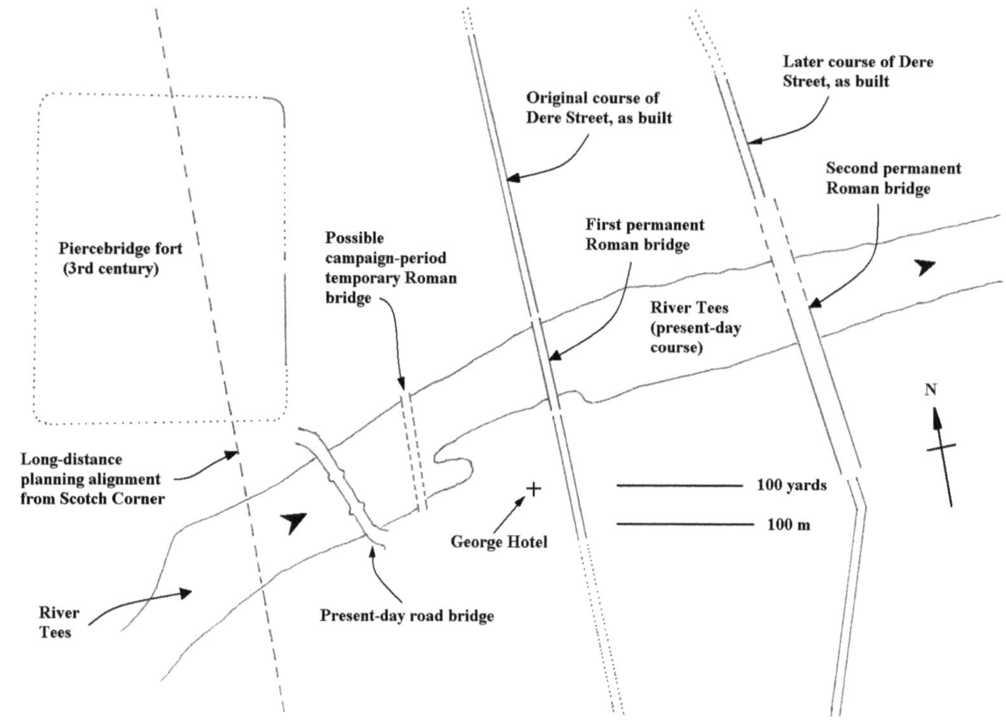

Figure 25 Approximate location and orientation of a possible campaign-period Roman bridge over the River Tees at Piercebridge

A further 30 to 35 miles (48 to 56 km) to the north, at Piercebridge beside the River Tees, a recent investigation by the Time Team television programme reported finding a timber structure a few yards (115 m) upstream of what has been known up to now as the first Roman timber bridge. The site of this latter well-known bridge lies at the back of the George Hotel, and it appears to have carried the first permanent course of Dere Street across the River Tees at that point. Figure 25 indicates the general position of the newly-discovered structure in relation to the known Roman bridge, as noted from the television presentation.[30] On the television programme, the structure was interpreted as an Iron Age bridge, but, to the author, a much more likely interpretation is that the structure was an early temporary Roman bridge, erected for its supply train to cross the river at that point in support of the campaigns further north.

It should be noted that the aforementioned first bridge to carry the permanent course of Dere Street over the Tees was probably not erected until the beginning of the second century,[31] so that the "temporary" bridge – whether Iron Age or Roman – may have had a life of some 30 to 40 years or more. Either way, this newly-discovered structure was only a few yards downstream of the projected long-distance alignment running north from Scotch Corner. Archaeologists have long suspected that an early Roman fort would have existed nearby, in order to protect this key crossing-point, but so far no firm evidence for such a fort has been detected. The known Roman fort at Piercebridge belongs to the third century but, as already noted, it lies exactly upon the projected long-distance planning line.

A fourth piece of evidence may lie at what is likely to have been the initial Roman crossing-place over the River Tyne, near Corbridge. Initially, the Romans appeared to have constructed their campaign base not at Corbridge but a mile or two (1.5 to 3 km) upstream at Red House farm. Only after 10 to 15 years or so did it seem that they relocated their Tyneside base to Corbridge, whereupon they then constructed the permanent course of Dere Street to run through that new site. Even so, the planning alignment that the author detected underlying the course of Dere Street from above Ebchester to the Tyne appears to arrive at the Tyne directly opposite the site of Red House farm, not Corbridge. See Figure 26.

It is difficult to press this last piece of somewhat circumstantial evidence very strongly, because the course of the River Tyne is believed to have altered considerably here since Roman times.[32] Nevertheless the alignment does appear to terminate opposite the base at Red House farm rather than at Corbridge, and the line of Dere Street, when built, does appear to curve from this alignment in order to cross the Tyne so as to run up to the later site. See Figure 27. Because of the importance of any earlier crossing to the Roman campaigns further north, its crossing-point can be expected to have been bridged whilst the base at Red House was in commission, but so far no physical indications of such a structure are understood to have been detected.

[30] The author is indebted to Dr. D. J. P. Mason for more precise information about the location of the new structure, but the orientation shown on Figure 25 has been recollected from what was shown on the television programme and should not be regarded as having been archaeologically verified.
[31] Poulter, 2010, 42-3

[32] For details of the changing course of this stretch of the River Tyne, see Snape, 2014, and Figure 2 in particular.

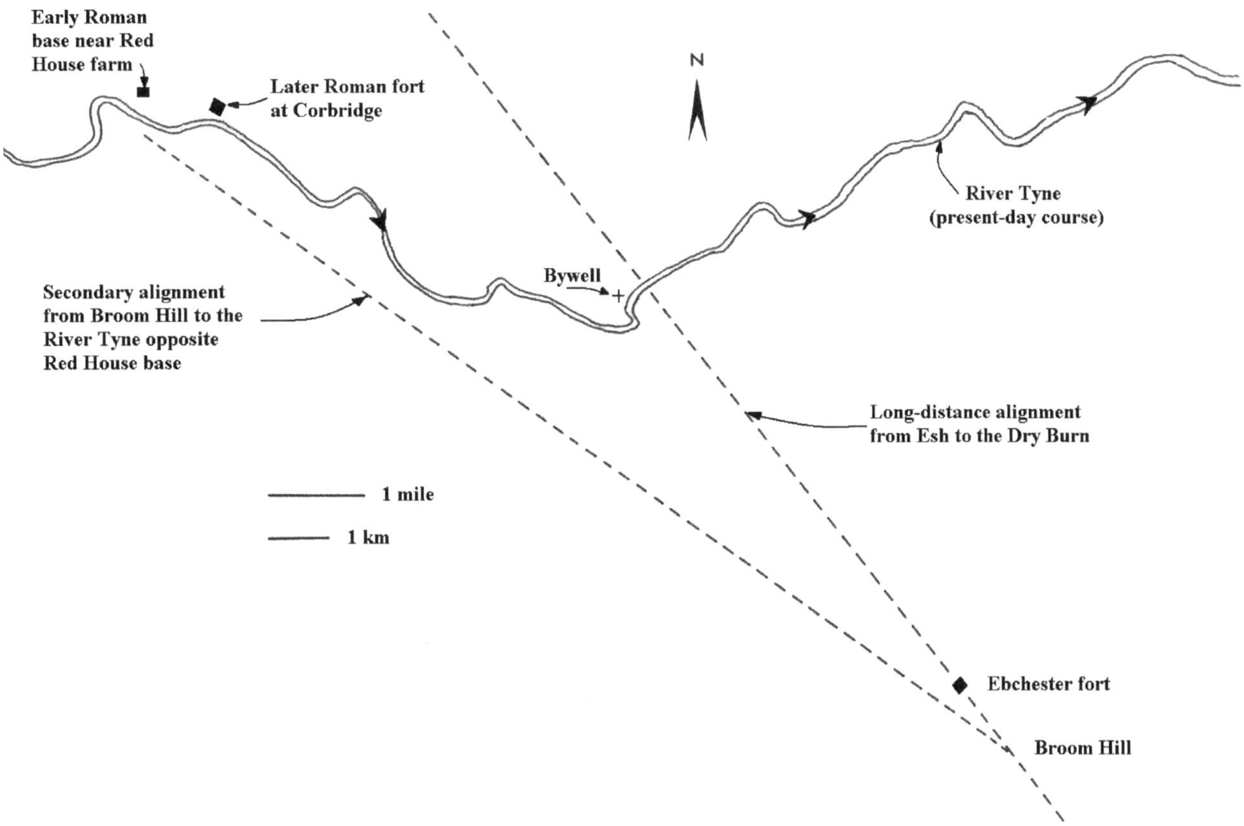

Figure 26 Arrangement of apparent planning alignments approaching the crossing of the River Tyne opposite the early Roman base at Red House farm

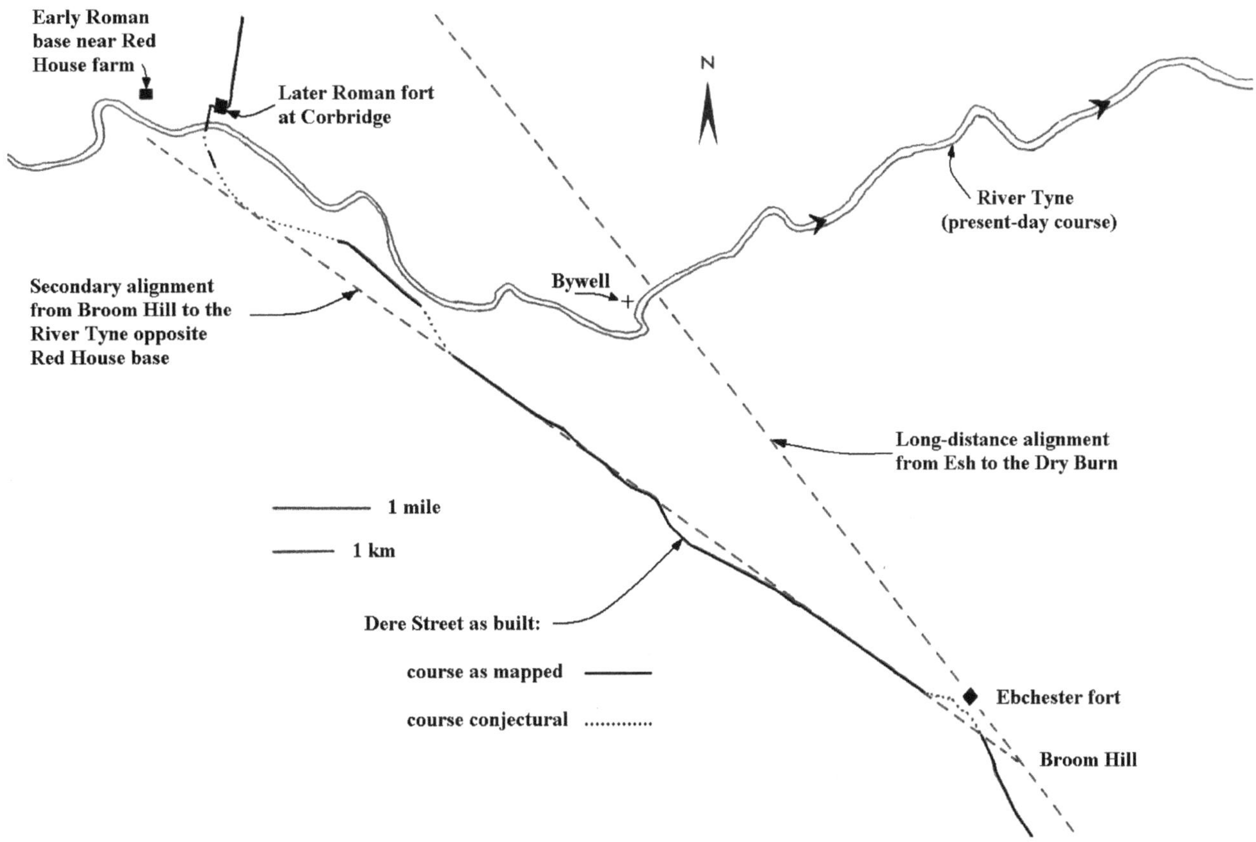

Figure 27 Course of Dere Street as built in relation to the apparent planning alignments, but crossing the Tyne opposite Roman Corbridge rather than Red House farm

The evidence for such short-lived Roman forts, and other installations belonging to the early years of their campaigns, is always likely to be difficult to find. It should be noted that at least two of the items of evidence described above were discovered fortuitously. Firstly, the fort at Roecliffe was discovered when a bypass around Boroughbridge was being laid out for the A1 trunk road. Secondly, it is understood that the earlier bridge structure at Piercebridge was first detected as a result of diving activities in the bed of the River Tees in connection with the known Roman bridge at the back of the George Hotel. In addition, although the late John Gillam had predicted its possible existence a few months beforehand,[33] the presence of a Roman base at Red House farm was only finally confirmed as a result of setting out a new course for the A69 trunk road between Newcastle and Carlisle. Only a few of these remains of early Roman activity had been suspected previously, therefore, and it can be imagined that many more examples of Roman activities, especially from the early conquest period, remain to be discovered and thus in due course cast further light upon what may have been happening at that time.

To conclude the case for this first possible reason for creating long-distance alignments: as far as the author is aware, there is no Roman description of the temporary planning and logistical-support activities proposed above, but, as has already been noted, neither is there a description of exactly how the *groma* may have been used for long-distance planning. There are, in Roman literature, accounts of engineers being sent out in advance of an army to clear and prepare the way before the main bodies of troops arrived to pass through the area. However, this may well be a different thing. The function of an army on campaign is to seek out and eradicate hostile forces, wherever they might be, and this focus is likely to determine the directions that the army would choose to take when in the field. It would only be after the conquest had been completed, and the next advance planned, when it would seem most likely that the supply and communication lines for the new campaign would be set out over the newly-conquered landscape. The evident need for such planning, on occasion, and the evidence presented above, make it seem quite likely to the author that long-distance alignments would sometimes have been created for this purpose.

A second possible reason for creating long-distance alignments

Long-distance alignments could have arisen from the Roman practice of surveying landscapes into square or rectangular grids, a process known as centuriation. Typically, these grids might be about 776 yards (710m) square, and this time descriptions of the Roman activities are recorded in their surviving literature.

The scholar most closely connected with research into such surveying activities in this country is John Peterson, and he has produced a useful glossary of the terms that the Roman surveyors employed when carrying out this work.[34] He has also produced numerous articles about possible examples of centuriation in Britain. Professor Sheppard Frere has also published a possible example of a centuriated landscape across northern Norfolk.[35] In this case the centuriation was interpreted as the result of a replanning of the confiscated landscape in the immediate aftermath of the Boudiccan revolt. There are some archaeologists who are sceptical about the presence of centuriated landscapes in Britain, but since such grids are well known in northern Africa and in Italy, and have also now been identified in Gaul and elsewhere in the Roman empire, there appears to the author to be no logical reason why Britain should be regarded as having been an exception.

Examples of some of the centuriation schemes that John Peterson has proposed in Britain are shown in Figure 28. It can be seen that they do not cover all of the country, and that they do not possess a common orientation. However, where they were created, they would have provided the Roman planners with the equivalent of our modern-day Ordnance Survey national grid, albeit that they would not have been aligned with the cardinal points of the compass.

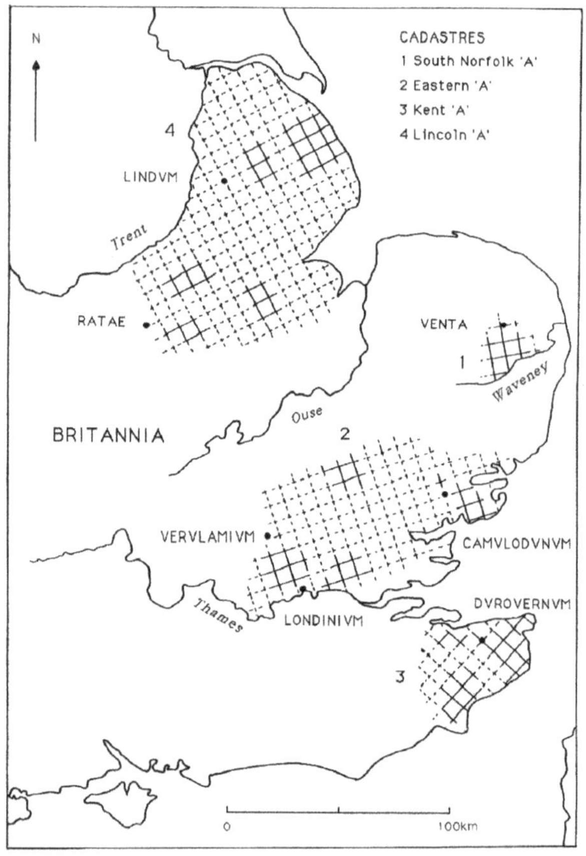

Figure 28 Schemes of centuriation identified in Britain by Dr John Peterson.
(Map reproduced by kind permission of Dr Peterson.)

[33] David Breeze, *pers. comm.*

[34] Peterson, 2009

[35] Frere, 2000

Centuriation is often thought of as having been a way of dividing up land for agricultural purposes – such as, for instance, allotting land to veterans on their discharge from the Roman army. However, John Peterson has made the important point that such examples of centuriation may also have been used by the Roman administration for determining the positions of forts, settlements, and roads. Indeed, in the formative days following a period of conquest, it may be suspected that this could have been the primary purpose of such schemes. John Peterson has demonstrated, via statistical analysis, the probability that Sextus Julius Frontinus, when governor of Britain during the second attempt to establish Roman control in south and central Wales, may have used such a scheme to determine the locations of forts and roads across the newly-conquered territory.[36] Thus this second possible reason for the creation of long-distance alignments would appear to fall within the third phase of the scenario outlined at the beginning of this section, i.e. to the period of consolidation after campaigning had been completed.

It should be noted, though, that there is no mention in the Roman literature about the use of centuriation for this purpose. This may be because, as John Peterson has pointed out, although there are examples of Roman literature on surveying which do date from the first and second centuries AD, much of the surviving Roman literature which is concerned with surveying comes from the later Roman period.[37] By that time, the planning of the locations of forts, settlements, and roads would largely have been things of the past, and the concerns of the surveyors would have become focussed upon the settlement of land disputes. This is indeed what the surviving Roman literature appears to reflect.[38]

The point should also be made that John Peterson is very rigorous in his assessments of the alignments and details of the evidence which he will accept for possible centuriation schemes. For instance, even a change in alignment of a fraction of a degree would cause him to question if a perceived parallel alignment might simply be a coincidence. Similarly, with regard to the proposed scheme of centuriation in northern Norfolk, Professor Frere noted that some of the alignments on the eastern side of the grid appeared to be as much as 1° out, and suggested an explanation for how this might have occurred. The important point here is not so much that an apparent Roman surveying error needed an explanation but that Professor Frere felt the need to offer one.

The work of John Peterson and Sheppard Frere reveals an appreciation of the remarkable degrees of precision which Roman surveyors were able to achieve routinely over very long distances. For instance, when examining the evidence for a probable scheme of centuriation in the Saône plain, north of Lyon in France, John Peterson found that a Roman road had been located (seemingly as part of the centuriation exercise) some 25km from one of the apparent survey lines to an accuracy of better than 10m, and that this appeared to reflect the degree of precision achieved by the surveyors throughout the rest of the scheme.[39]

Some of the schemes of centuriation could be very extensive, too. John Peterson reports that a scheme of centuriation in southern Tunisia covered an area of at least 68 by 133 miles (110 km by 215 km).[40] Thus some of the alignments of such a scheme of centuriation would have extended to more than 100 miles (160 km) in length. This needs to be borne in mind when attempting to assess the intentions of the Romans from the evidence which has come down to us. The Roman surveyors could, and evidently did at times, set out straight alignments – and alignments at right angles – with great accuracy over very long distances, and this is something that they would have needed to do when laying out any sizeable scheme of centuriation across the landscapes of Britain.

A third possible reason for creating long-distance alignments

However, extending a scheme of centuriation over a substantial tract of territory would not have been a quick process. John Peterson has estimated that to have conducted an outline survey for a scheme of centuriation stretching, say, over the 40 or so miles (64 km) from Thurrock in Essex to Great Chesterford in Cambridgeshire could have occupied up to two and a half years.[41] It may be imagined that Roman military planners, seeking to establish a network of forts and roads so as to maintain their grip upon newly-conquered territory, would have wanted decision-making information more rapidly than that. Hence the proposition – developed in correspondence between the author and Rob Entwistle – that the alignments that the latter had detected may have been the relics of survey lines set out to explore and report back on the topography and demography of the countryside in order to enable speedy decisions to be made.

Hence the systems model for this possible reason may be construed as follows. The Roman military authorities, wishing to make centralised decisions about the dispositions of a network of forts and roads in newly-conquered territory, would send out surveyors to report back on the topography and demography of the territory. These surveys would need to have been carried out along alignments which could be reproduced at the military headquarters, from which the decisions about where to site the forts and roads would then be relayed back to the commanders on the ground, most likely by means of the same survey lines which the surveyors had established.

[36] Peterson, 1995

[37] John Peterson *pers. comm.*

[38] See Campbell, 2000

[39] Peterson, 1993, 55-6

[40] Peterson, 1993, 242

[41] John Peterson *pers. comm.* 9 July 2011 to Rob Entwistle, also copied to the author.

Since this model, as far as the author is aware, is a novel one and it is not recorded in Roman literature, it will be worth exploring the arguments for it in more detail.

Argument 1: the occurrence – in some periods at least – of centralised Roman planning of its network of forts, settlements and roads

Category: presumption

It is often declared that the Roman administration was extremely bureaucratic, and, in the north of England we may get a glimpse of this amongst the writing tablets which have been found at the forts of Vindolanda and Carlisle in recent years.[42] In his *Agricola*, Tacitus states that Gnaeus Julius Agricola, when governor of Britain, determined the sites of his forts personally. If this statement should have been a fact rather than merely a conventional accolade for a praiseworthy governor, there may be an inference here that Agricola selected his sites when on campaign. Certainly, the chain of Roman forts continuing up from the River Tay into Aberdeenshire do give the impression of following a campaign trail, as indicated by the line of marching camps which seemingly reflect the progress of Agricola's forces further north up the eastern shoulder of Scotland.

However, some governors, such as Agricola's immediate predecessor Frontinus, and possibly even Petillius Cerealis who was governor of Britain before Frontinus, may have preferred to work remotely, on the information received from surveyors. As already mentioned, John Peterson has produced a case for the use of centuriation by Frontinus when planning a network of forts and roads in south and central Wales. In northern Britain, there may have been a similar but more urgent need for topographical and demographical information, especially if a large area of territory had been brought into the empire relatively quickly.

If so, this would have required the Roman administration to be able to record, in some form, the spatial distribution of the information being received from the surveyors. Effectively, this would have been equivalent to creating a modern-day map, even if its cartographic layout and conventions might have been different from what we would regard as a map today. The sorts of information required for decision-making would have been the descriptions of the landscape – the hills, valleys, marshes, the courses of major rivers, areas of woodland, existing tracks and routeways, etc. – and of the native populations – their numbers, allegiances, settlements, degrees of hostility, languages, customs, types of agriculture, and natures of their leaders or rulers, etc.

The presumption, then, is that, at certain times at least:

a) the Roman administration may have chosen to determine centrally (i.e. at its headquarters rather than out in the field) the disposition of the network of forts and roads to be established so as to maintain its grip over newly-conquered territory, and that

b) in order to do this, the Roman administration would have required a means of collecting reliable and spatially-related demographical and topographical information about the areas concerned, and upon which the decisions could be made.

Argument 2: that the lines of surveys would need to have been laid out at set angles

Category: proposition

Although it would have provided a coarser body of information than that which might have been obtainable from a scheme of centuriation, a set of rapidly-conducted and geometrically-related surveys radiating from one or more key points might have allowed the Roman administration to gain sufficient information for the purposes of their decision-making. For instance, such lines could be set out from one or more coastal points to explore the interior of the country and also the territory to the north and south of the chosen points. With reference to Figure 19, it is possible that those lines apparently radiating from Lancaster could be an example of this.

The need for the survey lines to be geometrically-related would have been so that the spatial relationships of the information received back from the surveys could be reproduced remotely at the headquarters of the Roman administration. Since Roman surveyors did not measure angles in degrees[43] – and in any case did not have the instruments to do so – they would need to have set out the survey lines at angles which could be easily and reliably reproduced both on the ground and at the administrative headquarters. Such angles would include right angles, 45° angles, and the angles of equilateral and Pythagorean triangles.

It might be thought that it would be important for such groups of survey lines to be related to the points of the compass, i.e. north-south, and east-west. However, as with the evidence for the examples of centuriation cited above, this does not in general seem have been important to the Romans. Instead, it is possible that some of the lines may have been set out to form links with existing Roman sites or with surveys conducted by other teams, possibly previously.

[42] For a depiction of the extent of Roman bureaucracy, see Breeze, 2011, 183. For even more detail, see Austin and Rankov, 1995, 155-61.

[43] Lewis, 2001, 41

For instance, the central of the three survey lines that apparently radiated from Lancaster may have been extended across the Pennines to link up with Roman surveys on the eastern side of the country. See Figure 19. It is also conceivable that the southern line from Lancaster may have been put out as a "feeler" towards an already-established Roman fort at Manchester, or else to the survey which may have helped to determine that fort's location.

It may be noted that if the intention had indeed been to aim for Manchester, then this would have represented good shooting on the part of the Roman surveyors. The two sites are decidedly not intervisible. However, if the line should have run exactly between Lancaster and Manchester, it would have suggested that the bearing between the two locations would already have been established by some prior form of survey, such as via centuriation. The fact that the alignment runs close to but not exactly between the two Roman sites suggests that their locations were the results of separate survey activities.

The proposition, then, is that for rapid topographical and demographical surveys of new or unexplored territory, the Romans could have laid out groups of survey lines across the landscapes at the kinds of angles on the ground which could then be easily reproduced at the Roman administrative headquarters.

Argument 3: the need for the survey lines to be straight

Category: proposition

If the survey lines had been set out across the territory with geometrical relationships to each other, in order that the information reported from the surveys could be reliably recorded spatially at the administrative headquarters, then it would have been important that the geometrical relationships should be preserved as the survey lines were extended. Hence it would have been important that the survey lines should be laid out as straight as possible.

In practice, an adherence to perfect straightness might not have been absolutely vital for the purposes of the surveys, provided that the teams setting out the alignments should not have been consistently veering off line in the same direction at each sighting point. In this case their errors would have been cumulative. However, as it happens, setting out such straight alignments would have been readily accomplished with instruments such as the *groma* or the *dioptra*, especially with the precision which might be achievable by surveying with single points of light at night.

Hence the proposition is that the survey lines were kept as straight as possible in order to preserve the geometry of the relationships between the alignments, and that this would have been easy for the Romans to have achieved with the surveying instruments which they were using.

Argument 4: the surveying process

Category: deduction

It appears likely that the Roman surveyors would thus have extended their survey lines across the countryside from high point to high point, simply by sighting backwards to the previous high point and onwards to the next. Surveying in this way at night would have allowed considerable distances to be covered in only a few days, regardless of the nature of the landscape in between. The topographical and demographical surveying of the territory in between the high points would have been undertaken in daylight, of course, and may be considered likely to have occupied by far the greater parts of the surveyors' time.

Argument 5: the decision-making and instruction process

Category: deduction

Once the surveyors' reports had been received and recorded at the administrative headquarters, decisions could then be made about the positions of the forts, and the roads to be constructed between them. Several factors would need to be considered. For instance, the forts would not only need to cover key points but be one day's march apart, once the roads had been built.

These decisions would need to be relayed to the commanders on the ground, and the obvious way in which to do this would have been to issue instructions to the commanders in relation to the survey lines which had already been established and marked out across the landscapes. Indeed, in many cases, these lines might have offered the only reliable frames of reference.

Many of the locations for the forts would, of course, need to have been sited well away from the survey lines. Indeed, some of the locations would probably have been identified as possible sites for forts by the surveyors during the courses of their surveys. Even when it should have been decided to locate forts on or close by the survey lines, however, the commanders on the ground would need to have been given some latitude in implementing the decisions from headquarters, i.e. in selecting, in their opinions, the most suitable sites for the forts in relation to their instructions.

Hence, even when an order might have been issued to erect a fort on or near one of the survey lines, not every suitable site would have fallen exactly upon such a line. Thus, as discussed in Section 7, the fort at Kirkby Thore lies some 300 to 400 yards (275 to 365 m) west of the projected survey line from Lancaster to High Rochester, on a site overlooking the Trout Beck near its junction with the River Eden. See Figure 19. Similarly, the fort at Whitley Castle lies some 600 to 700 yards (550 to 640 m) to the west of this same line, on a knoll above the South Tyne River which would have offered a strong point for defence and observation of that area. On the other hand, in certain cases, suitable locations for fort sites might have been found to lie exactly upon such a survey line. This may have occurred with the forts at Kirkham and at Elslack, on the apparent alignment to Aldborough which is shown on Figure 19 and which is described in more detail in Section 8. The fact that this positioning of forts exactly upon apparent survey lines seems to occur in other places as well – for instance, as already noted, along Dere Street – might be taken as some indication of the degree of control exercised by the Roman administration in this process.

The model as thus conceived is therefore based upon one presumption, two propositions, and two deductions. It is crucially dependent upon the first of these – the presumption that at certain times at least, the Roman administration would have sought to plan their network of forts and roads remotely, at headquarters, rather than in the field. Given this presumption, the model is considered to have been an entirely workable one.

However, aside from the evidence on the ground, which will be examined in detail in Sections 7 to 9, it was wondered if there should be other indications that the proposed model might not just be feasible but one which might have actually been employed by surveyors in practice. Hence the model, which had essentially been developed in the autumn of 2010, was held back for 2 to 3 years before being submitted for publication in this monograph. In fact, the reasons for holding publication back were twofold: (a) to see if, on reflection, some flaw might be spotted in the arguments, and (b) to see if instances of the application of the model might be observed elsewhere. In the event, no flaws were detected in the arguments, but two endorsements were received of the practicality of the model.

The first of these came at a meeting in 2011 with Dave Robson, who, as already mentioned, had had experience of laying out new roads in northern Nigeria. The main purpose of the meeting had been to examine a network of Roman roads which Dave Robson believed he had identified across County Durham. However, as an aside from the main discussion, the opportunity was taken to ask Mr Robson, as a surveyor with considerable practical experience, what he thought of the notion of attempting to execute rapid surveys of unknown territory by following alignments radiating at known angles from key points, such as Rob Entwistle appeared to have detected at Lancaster. Dave Robson's response was immediate and positive: yes, he could certainly see that a process such as that would be both a rapid and effective means of surveying unknown territory and its populations. This is the second instance on which the author has been indebted to Mr Robson for his practical experience and suggestions.

The second endorsement came a few months later, when the author's attention was drawn to a report made by Jefferson Davis, Secretary of State for War, to the President of the United States of America in 1853.[44] At that time, the USA, still largely based upon the eastern side of the continent, was concerned about how it would be able to defend itself against attacks on its southern border and along its Pacific coast. Moving troops rapidly into those areas from the eastern or central USA was considered likely to prove difficult if not virtually impossible because it was known that there were extensive tracts of desert to be crossed, with no water or fodder available for the draft animals that would be required to move the transports.

It was considered that constructing one or more railroads would offer a partial solution to this need, but the question was: from where to where, and through what territory? Although numerous explorers and settlers had by then passed through the territories concerned, the accounts that the US government possessed about the landscapes and the native Americans which occupied them were felt to be contradictory and unreliable. What the US government needed was reliable information upon which decisions could be made. These were primarily about where one or more railroads might be constructed but also about ancillary factors such as the potentials for settlement, and the exploitation of mineral resources. Hence a series of surveys were commissioned to assess various territories across the western and southern parts of North America and report back in one year. The following is a distillation of the essence of Jefferson Davis's report.

[44] Davis, 1853

Analysis of that part of the Report relating to explorations and surveys, submitted by Jefferson Davis, Secretary of State for War, to the President of the United States, dated 1st December, 1853

Driver:

the need to be able to defend the southern border and western territory of the United States, and its Pacific coast, in (a) the absence of navigable streams in a large portion of the territory and (b) the presence of extensive areas of desert which are impassable for draft animals because of the absence of grass or water. This renders the transport of munitions and supplies extremely slow and expensive and hinders operations against native Americans to the point where the operations can become ineffective. A railroad would only partially remove the difficulty, but it would be able to transport troops and supply depots along the route and at the extremity of the line.

Requirement:

to ascertain the most practical and economical route for the construction and operation of a railroad from the Mississippi river to the Pacific basin, across an area of which existing knowledge is patchy and reports of its form and nature are frequently contradictory.

Problem:

to do so within a very limited timescale and budget, neither of which would be sufficient for a complete solution (i.e. survey) across such a vast area.

Objectives set:

1) to examine the suitability of the land profile for the construction of a railway, together with the fitness of rock for use in masonry and the availability of water for engines and irrigation

2) to examine the capacity of regions to support a human population and furnish articles of commerce.

Process:

separate parties of surveyors to explore from the Mississippi to the Pacific coast (i.e. east to west) in the vicinities of specified latitudes, namely the 47th, 38th, 35th, and 32nd parallels, together with a variety of surveys from west to east, starting from the Pacific coast, to seek potential passes through the mountain ranges on the western side of the territory (see Figure 29); all parties to report back their findings for assessment

the parties to be accompanied by small numbers of selected troops to provide cover from any hostile attentions received during the process.

Instructions:

1) to observe and note all the objects and phenomena which would have an immediate or remote bearing upon the railroad, or which may serve to develop the resources, peculiarities and climate of the country

2) to determine variations in the direction and intensity of the magnetic force across the territory

3) to observe the prevailing wind, amounts of rain and snow, and the temperature and humidity of the atmosphere

4) to report on the geology of the country, and to collect specimens of rocks and soils, plants and animals

5) to accumulate information about the native American tribes in the regions traversed, including their locations, characters, numbers, histories, traditions, languages and customs.

Commentary:

throughout, the report seems much more concerned about the costs of the surveying exercise and the nature of the territory to be covered than it does about hostile actions from the natives. Nevertheless all the parties seem to have been equipped with small escorts of picked troops, so the threat appears to have been recognised even if not regarded as a very substantial one.

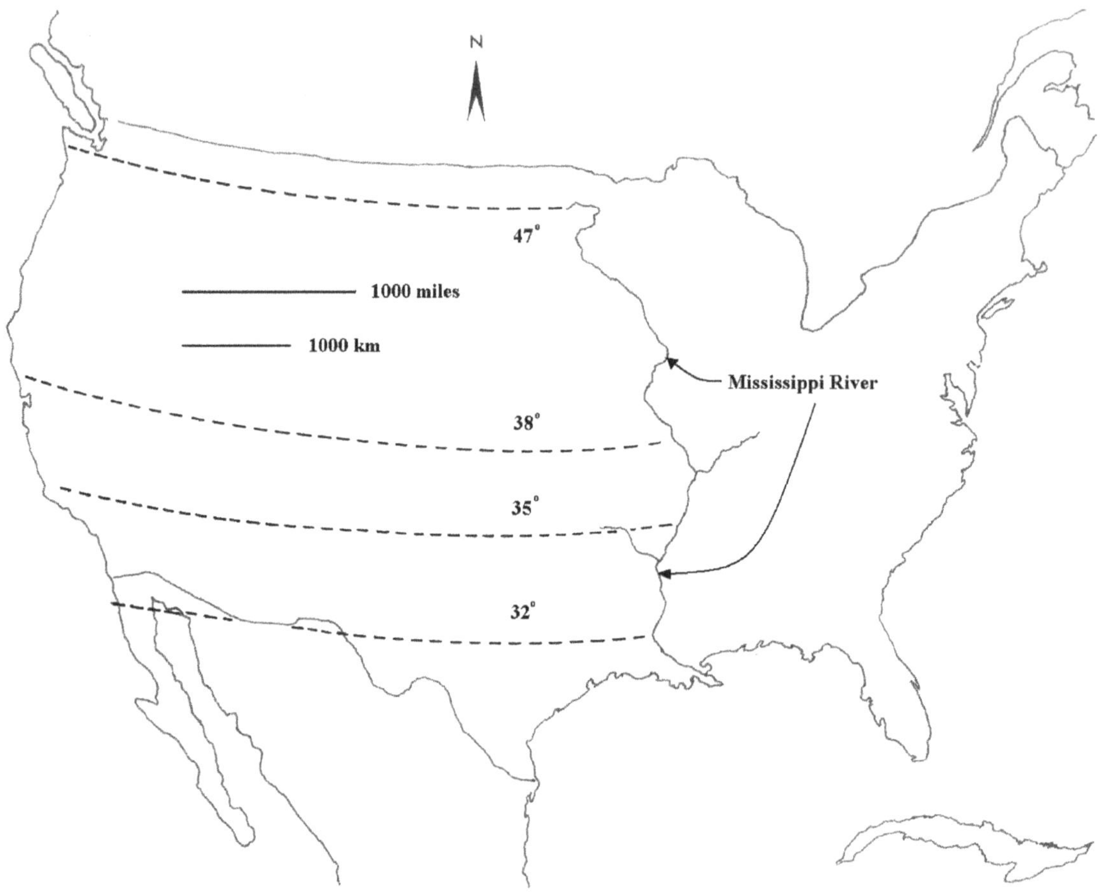

Figure 29 Indications of the extent of the four main explorations and surveys commissioned by the military authorities of the USA in the early 1850s, in connection with the possible construction of a railroad to transport troops and munitions to its southern and western borderlands. The four main survey parties were directed to explore from the Mississippi to the Pacific coast (i.e. east to west) in the vicinities of the 47th, 38th, 35th, and 32nd parallels

Apart from the second instruction, which was concerned with measuring magnetic force, it is considered that in principle this report records an almost exact application of the model proposed above. The chief differences are that, by 1853, (a) an internationally-agreed global positioning system of latitudes and longitudes had been developed, and (b) that the US surveyors possessed the instruments with which to determine where they were in relation to it. Without such a grid, the Romans would have had to create their own geometry. It is therefore considered that this practical application of the model in similar circumstances to those which are likely to have faced the Roman army – i.e. where to deploy its military infrastructure across an inadequately known territory – does offer strong support for the proposition that the Romans could well have applied such tactics some 1700 to 1800 years ago in northern England, and possibly elsewhere in Britain too.

A fourth possible reason for creating long-distance alignments

Finally, there is the possibility that the planners of Roman roads, when commissioned to lay out a course between two or more points, may in certain cases not have been able to benefit from the existence of any previously-created survey lines or grids upon which to work. In such instances it is possible that they may have set out some of the long-distance alignments themselves, so as to provide guidance for the overall direction that their road should take.

It is believed that occurrences of this may be discerned with each of the three roads which are described in sections 7 to 9. In addition, this appears to have been the way in which the entire course of Hadrian's Wall was set out, and also that of the Vallum which runs behind it.

Finally, a few weeks after the draft of this monograph had been written, but without having had any sight of its contents, Mike Haken of the Yorkshire Archaeological Society had sent his observations to the author about the planning of the Roman road (Margary 28) which runs for some 70 to 75 miles (112 to 120 km) from Ermine Street, north of Lincoln, to York via Bawtry, Doncaster, and Castleford. This road thus provided the Romans with a route between Lincoln and York which avoided a crossing of the Humber. Although the author has not examined the course of this road on the ground, from Mike Haken's analysis of its planning it would appear to fit well within this fourth category, in which long-distance alignments would be set out first and then the actual course of the road on the ground would then be worked out from them.[45] It was also interesting to find that, working entirely independently, Mike Haken had reached conclusions about Roman planning processes which were similar to those which Rob Entwistle and the author had developed some two or three years earlier.

Concluding discussion

The foregoing propositions about Roman planning processes raise issues which merit a brief examination of what is currently known about Roman surveying, map-making, and about the information which would have been available to Roman governors and generals.

Firstly, as already noted, much of the surviving Roman literature about surveying is concerned with the measurement of parcels of land in connection with property rights, and much of the modern literature about Roman surveying adopts a similar focus. For instance, in Dilke's contribution to *Aufstieg und Niedergang der römischen Welt*, "land surveys" appear to be taken throughout to mean "centuriation".[46] Dilke reports that the results of such surveys would be drawn up in two copies, one of which would be retained locally whilst the other would be lodged in the emperor's record office in Rome, the *tabularium*, built in 78 BC.[47] Professor Brian Campbell adds that measurements, and lists of settlers where appropriate, could be stored there in addition to the plans of field systems.[48] It is clear that the office in Rome provided the ultimate source of information when it came to arguments about land division and previous decisions.

This definition of the word "surveying" is, however, much more limited in scope than the meaning of the word which is intended in this monograph, in which entire landscapes would be assessed both topographically and demographically with a view to their occupation and administration. It is clear that the Romans did undertake wide-scale surveys of this kind from time to time. Sherk describes fourteen exploratory Roman military expeditions to territories which brought back not only geographical details but also information about the natures of the peoples in them, their agricultures, and their cities and the distances between them.[49] Sherk adds that his list is not exhaustive.

It appears likely that maps were used by at least two of these expeditions to illustrate the results of their surveys. In the case of a survey of Ethiopia under the Emperor Nero, the report apparently recorded the exact distances between settlements and staging-posts, and showed the locations of palm groves.[50] This would seem to have required some form of map, whilst the report of the campaigns of Domitius Corbulo in Armenia in AD 57-8 refers to "topographical depictions" (*situs depicti*) which again suggests the presentation of the information in cartographical form.[51]

Dilke states that the Romans used maps for various purposes, which he lists as:

- world maps
- maps illustrating geographical treatises
- road maps and itineraries
- official and military maps
- detailed town plans
- surveyors' maps

and he adds that other technical maps and plans are likely to have existed but have not survived.[52]

With regard to road maps and itineraries, Austin and Rankov note that, for general purposes, the Roman army, and Romans in general, seem to have been accustomed to make use of them in the form of illustrated strips, or even just as bare lists of road stations – the *itinereria*.[53] Once the roads had been built, this would have been a very natural way of presenting the most relevant information required by people travelling along them. Ogilby used this style of mapping in his *Britannia* atlas,[54] and Taylor and Skinner used very much the same technique when publishing their *Road Atlas of North Britain or Scotland* in 1776.[55] This latter atlas shows many of the military and other roads of Scotland, with every milestone numbered, the locations of suitable inns and way-stations, and with nearby hills, rivers, and lochs depicted in somewhat schematic form on either side. But, before any roads existed, such maps would have been of no use when planning where the roads should be set out across the landscape. Much wider topographical surveys would have been required for that, with more accurate cartographical representations of the dispositions of the physical features and settlements upon the landscape.

It is known that the Romans could achieve this accurate cartographical representation via the process of centuriation, because the location of every feature on a centuriated landscape could be specified precisely in terms of the orthogonal coordinates of the centuriation. There would seem to be no reason why the Romans should

[45] It is anticipated that Mike Haken's analysis of the planning of this road will be published in the forthcoming months. See Haken, 2014/15.
[46] Dilke, 1974
[47] Dilke, 1971, 113
[48] Brian Campbell, *pers. comm.*
[49] Sherk, 1974, 537-43
[50] Austin and Rankov, 1995, 31
[51] Austin and Rankov, 1995, 114
[52] Dilke, 1971, 109
[53] Austin and Rankov, 1995, 115
[54] Ogilby, 1675
[55] Taylor and Skinner, 1776

not have been able to do the same where the coordinates would not have been orthogonal but would have possessed simple geometrical relationships such as being at readily-constructed angles to each other.

With regard to the information which would have been available to Roman governors and generals, Austin and Rankov do discuss this in considerable detail.[56] However, the information that they describe is almost entirely concerned with the military and political intelligence required in anticipation of hostilities, and during the execution of those hostilities, rather than with the information needed for the processes of consolidation after a conquest.

With regard to where such military and political information might have been held, even if it should have been obtained by means of a survey, it may be considered unlikely that such intelligence would have been kept at the *tabularium* in Rome. Firstly, such information would be transient, and secondly, it could well be regarded as confidential at the time. It would appear much more likely that such records would have been kept and maintained by the governors, their staff, and the generals in the provinces, which is where it would have been needed. David Breeze indicates the kind of archive which such officials could have kept, and the large staff of adjutants and clerks which would have been required to manage it.[57]

[56] Austin and Rankov, 1995

[57] Breeze, 2011, 183

Section 6: Examination of possible processes for converting parts of the long-distance alignments into roads

Although the teams of surveyors and engineers setting out the courses of Roman roads may have been acting upon instructions from higher authority, it is believed that when those planning teams worked out the final lines for their roads, they were doing so on the ground. This belief is founded upon the author's experience over more than 50 years of following and studying the courses of Roman roads across numerous landscapes, and especially those in northern Britain. Contrary to this view, Dr. Hugh Davies has proposed that not only were the general directions of Roman roads planned centrally, but that the actual lines that the roads should take were also plotted centrally, i.e. on some form of map at the administrative headquarters. Dr. Davies has argued that these plotted lines would then be transferred to the actual landscape by scaling up the line on the ground.[58] It is considered that this proposition is not correct, because the evidence does not support it.[59]

Nevertheless, it can be taken that the road planners would always have been working under some form of instructions from higher authority, and the question is how prescriptive such instructions would have been, and how much would have been left to the discretion of the planning teams. No doubt the degrees of freedom allowed to the planners would have varied from time to time. In any case, the road planners would have had to find practical solutions to physical obstacles encountered along the way, such as small bogs, precipices, and minor ravines and river crossings. These would have prevented any overlying instructions from being followed exactly.

The preceding section presented three different possible ways in which long-distance alignments might have been created by Roman surveyors. This was in addition to the probability that in some cases the alignments may also have been set out by the road planners themselves. Each of the three surveying mechanisms would have left alignments across the landscapes which, if preserved, could have been used subsequently by the planning teams to help set out the courses of their roads on a permanent basis. It should be noted that sometimes, as appears to have been the case with Dere Street, there might have been considerable gaps in time between the creation of the original long-distance alignments and their later re-use for the purposes of setting out the courses of the permanent Roman roads that we know today.[60]

[58] Davies, 2002, 50-2

[59] Poulter, 2009, 4; 2010, 23-4. For a more detailed critique of Davies's proposal, see Lewis, 2001, 224-6.

[60] For Dere Street, see Poulter, 2009, 25-9; Poulter, 2010, 38-46.

Re-use of former campaign trails

As proposed in the preceding section, the first mechanism which might have created such long-distance Roman alignments could have arisen from the need to provide general directional guidance for the sappers tasked with engineering the routes for the supply trains and communications lines which would have been required in support of the Roman army campaigning ahead. Naturally, such alignments, running straight for several miles, would have been unlikely to intersect rivers at suitable crossing-places, nor to have followed exactly the same lines as existing trackways. However, by erecting bridges at the nearest suitable crossing-points to the alignments, and in adapting those parts of existing trackways which happened to conform with the broad general direction, the alignments would have helped to ensure that traffic could be channelled along the right lines. It can be expected, therefore, that any alignments created for this purpose would have been quite well marked out on the ground, probably with intervisible posts or flags, and these can be expected to have remained in position for at least as long as traffic continued to use the temporary arrangements.

It appears that this is what may have been the case with Roman Dere Street if, as indicated in Section 5, much of its course between the Vale of York and the Dry Burn, just north of the River Tyne, had indeed been based upon what had originally been a series of supply train alignments. When the southern part of the permanent course of Dere Street came to be built, some 15 or more years may have passed since the original long-distance alignment had been set out northwards from Tadcaster.[61] By that time, the origin for most of the traffic to the north can be expected to have shifted from Tadcaster (or Roall) to York, where a legionary fortress had been established by the Romans in the early AD 70s. Accordingly, the permanent course of Dere Street appears to have been set out west-northwestwards from York to cross the River Nidd at the earliest opportunity and then turn northwards to run above the flood plain of the River Ouse before crossing the River Ure at a new point, Aldborough. As shown on Figure 23, it was only at Dishforth, some 4½ miles (7 km) north of Aldborough, that the permanent course of the road seemed to pick up the original long-distance alignment and follow it to Healam Bridge and then continue along the succeeding alignment to Leeming.

Further south, another Roman road – the Rudgate – was set out south-westwards from the permanent course of Dere

[61] Or even from further south, in the vicinity of the Roman site of Roall, as discussed in Section 5.

Street at Whixley. This road seems to have been heading for a destination somewhat to the west of Tadcaster but, along its rather winding way, it appears to make brief use of the original long-distance alignment near Bickerton. See Figure 15. From the way that the Rudgate diverges from a straight stretch of Dere Street at Whixley, it is evident that it was set out later than the permanent course of Dere Street at this point. This implies that the markings of the original long-distance alignment would still have been traceable by that time, i.e. even after the permanent course of Dere Street had been set out. It may also imply that the original alignment could still have been in use to some extent as a supply line, as surmised in Section 5. In this case, as already suggested, the upper part of the Rudgate would represent a diversion from that original alignment so as to cross the Ure at the new crossing-point at Aldborough, rather than, as formerly, near Roecliffe.

North of Scotch Corner, however, the permanent course of Dere Street appears to have made only intermittent use of the original long-distance alignments which have been detected. See Figures 22 and 23. The longest stretch of apparent re-use of an alignment ran to the north of the Tyne, from Beukley to the Dry Burn, where, as far as can be seen, this particular series of alignments coming up from the south seems to have ended. See Figure 11. Conceivably, this may reflect the extent of Roman conquest at the time when the long-distance alignments were first set out.[62]

Use of centuriated grids

The second mechanism by which long-distance alignments may have been created was that of laying out a centuriated grid over a substantial area of territory. The existence of such a grid, marked out on the ground, would have made road planning relatively easy. For instance, if it had been decided to construct a road linking two points on the grid, A and B, as shown in Figure 30, then the bearing to be adopted would simply have been 1 across on the grid and 2 up. Moreover, the accuracy with which the bearing was being followed could be checked at each intersection with the grid.

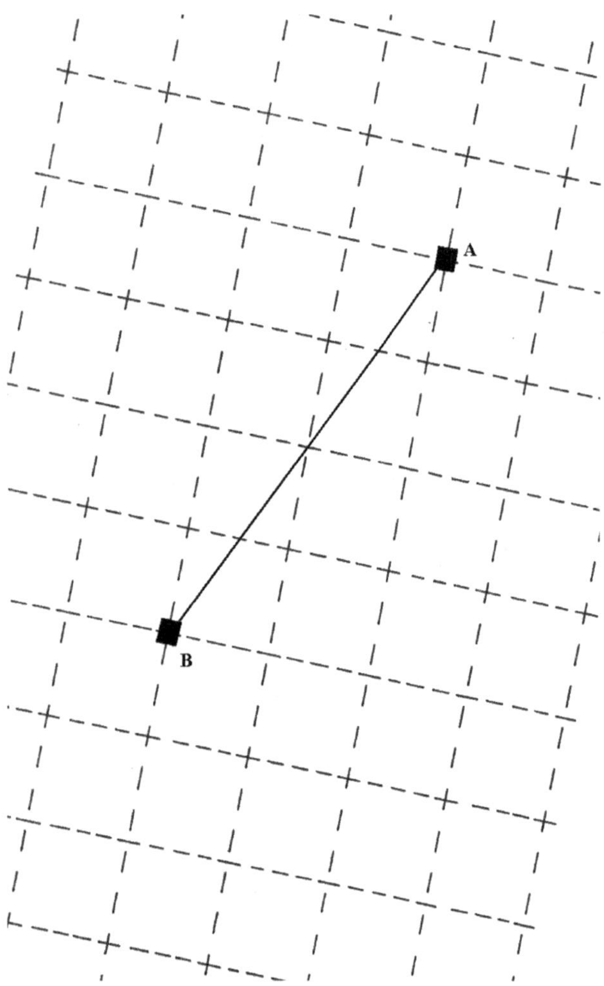

Figure 30 Using a centuriated grid to lay out the course of a road between two points located on the grid

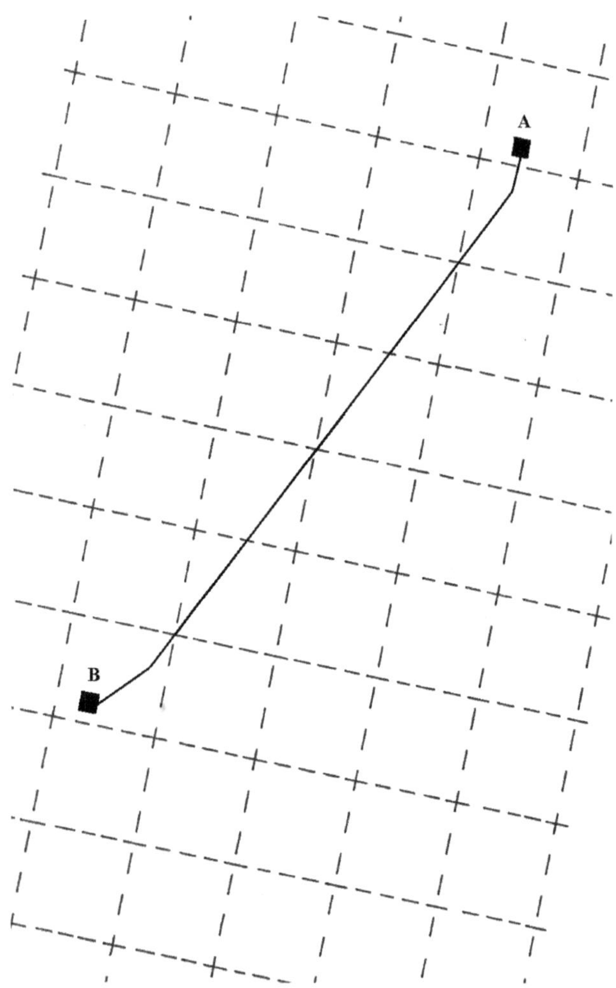

Figure 31 Using a centuriated grid to lay out the course of a road between two points not located on the grid, with minor adjustments to the line of the road at its ends in order to reach the final destinations

[62] However, see the discussion in Section 10 for an alternative possibility.

If any obstacles should be encountered when building the road between A and B, then a deviation could be inserted to circumvent the obstacle and then a return made to the original bearing. As already described, it appears that this would have been standard Roman practice. If A and/or B should happen not to lie exactly along the intersections or lines of the grid, however, then rather than attempt to vary the bearing, small adjustments to the alignment of the road could be made from and to the termini when they were in within viewing distance. See Figure 31. This often seems to have been what happened in practice. It would have allowed the accuracy of the main bearing to continue to be checked at its intersections with the grid, whereas this facility would have been forfeited if the bearing had been varied from the grid.

As John Peterson has pointed out, the use of centuriated grids may help to explain the phenomenon in which two stretches of Roman road run exactly parallel to each other but sometimes quite some distance apart. John Peterson calls this "bayonet form" of road, and it appears to occur more often than might be expected by chance. It appears to occur, for instance, in the second of the roads which are examined in detail in this monograph, in Section 8.

Two mechanisms might be responsible for such Roman roads being in "bayonet form". The first is where a diagonal bearing across the grid is displaced sideways for some reason, as shown in Figure 32. The second could be where two different roads happen to follow parts of the same grid lines, as shown in Figure 33, or else employ diagonals on the same bearing within the grid.

The occurrence of "bayonet form" in one or more roads does not necessarily prove that a centuriated landscape had existed, however. In principle, at least, parallel alignments could have been created by marking out survey lines at the same set angles from a common exploratory survey alignment, if both of the resulting survey lines had then been used to help create the courses of Roman roads. See Figure 34.

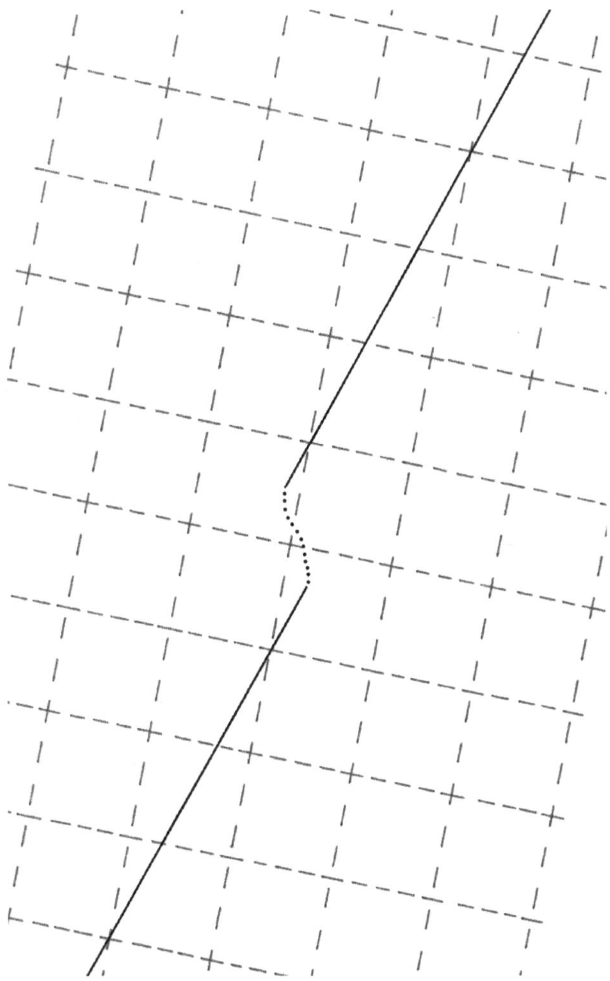

Figure 32 "Bayonet form" of Roman road caused by displacement of a road within a centuriated grid

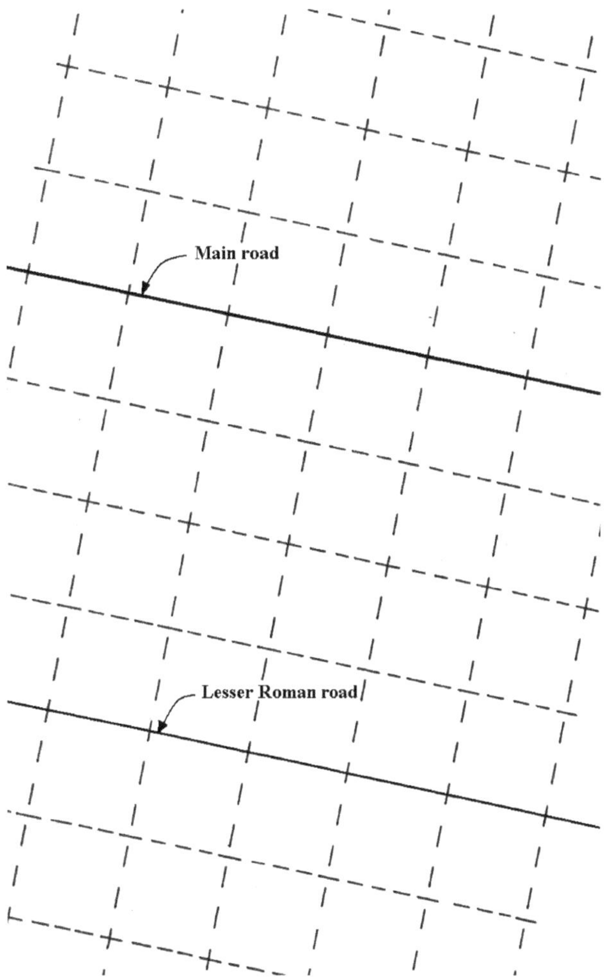

Figure 33 "Bayonet form" of Roman road caused by two separate roads making use of the same centuriated grid

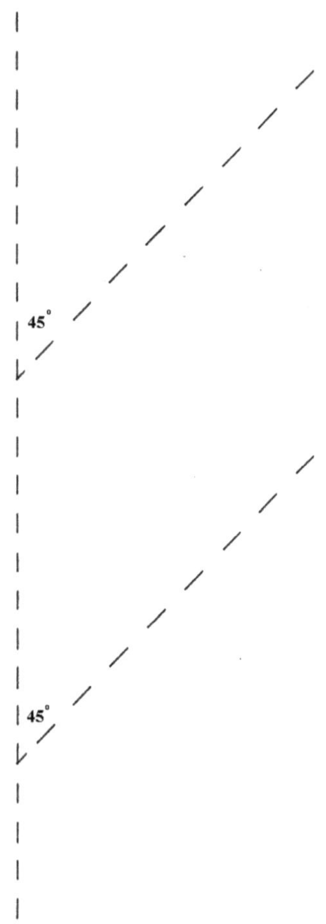

Figure 34 "Bayonet form" of Roman road caused by setting out survey lines at the same set angle from a common exploratory survey alignment

Use of exploratory survey lines

If the long-distance alignments had been created by the third possible mechanism, i.e. in order to conduct rapid surveys across unknown territories, and if the road planners had deemed it expedient – or conceivably had been instructed – to make use of parts of such alignments for the courses of their roads, they would still have had a lot of work to do. To begin with, when the long-distance alignments had been set out simply as survey lines, from high point to distant high point, it can be expected that there would have been no pressing need to mark out each alignment closely on the ground in between. It seems likely that the topographical and demographical surveys could have been carried out without making precise measurements of the position of every feature relative to the survey line. See Figure 35.

If the road, on the other hand, were to follow the surveyed alignment closely on the ground, the line would need to have been marked out much more closely on the ground as well. This is because there would almost inevitably have been several places across the intervening landscape from which one or both of the survey markers on the distant high points would have been invisible. See Figure 36. Thus, to follow the alignment exactly, the road planners would need to have extended the line on the ground, starting from one, or else possibly both, distant high points and working towards the other.

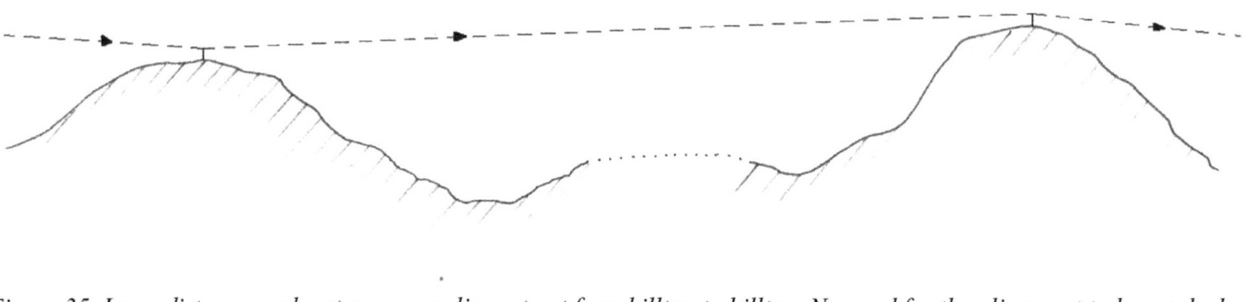

Figure 35 Long-distance exploratory survey line set out from hilltop to hilltop. No need for the alignment to be marked out on the ground in between

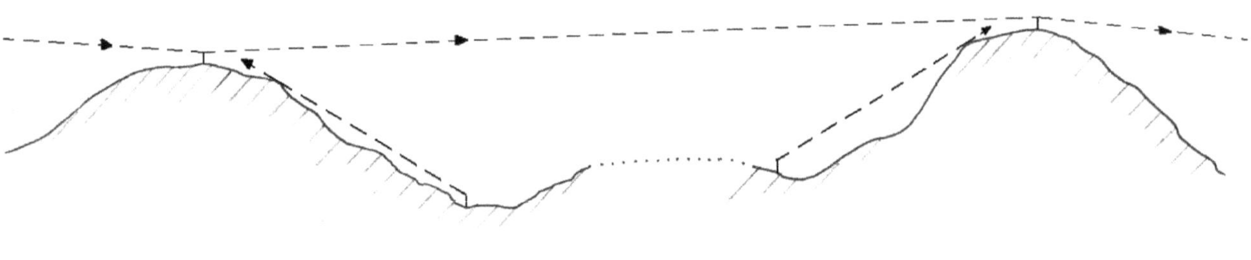

Figure 36 Long-distance survey points not always visible from places on the ground in between

Sometimes, depending upon the topography, these intermediary markers might have been only a few hundred yards or metres apart. However, it is doubtful that such markers would ever have been more than 1 to 2 miles (1.5 to 3 km) apart, even when the landscape might have been such that the road planners could see much further. This is because it can be taken that the road planners would have been working in daylight, whereas it seems probable that the original survey lines could have been set out over much longer distances at night. In daylight, the maximum distance at which visible instructions could be given to staff ahead – for instance, to move a little to the right or left – is likely to have been less than 2 miles (3 km), even in conditions of good visibility.

A consequence of these intermediate markings-out would have been an accumulation of small errors of alignment, some of which would have cancelled each other out. However the road planners would have been doing well if they should still have found themselves exactly on the original survey line when they finally arrived at the next high point – or, if they had been working towards each other from each high point, when they had sought to meet up in between.

In some places between the two distant high points, one or other of the high points might have come into view for the road planners. In such cases it should be noted that if the road planners should have been working towards a distant high point and had become a little off line, then sighting on to the distant high point ahead would enable them to converge on to the original alignment. On the other hand, if the road planners, when becoming a little off line, should only have been able to sight upon the distant high point to their rear then this would cause them to diverge further from the original alignment. This is illustrated in Figures 37 and 38.

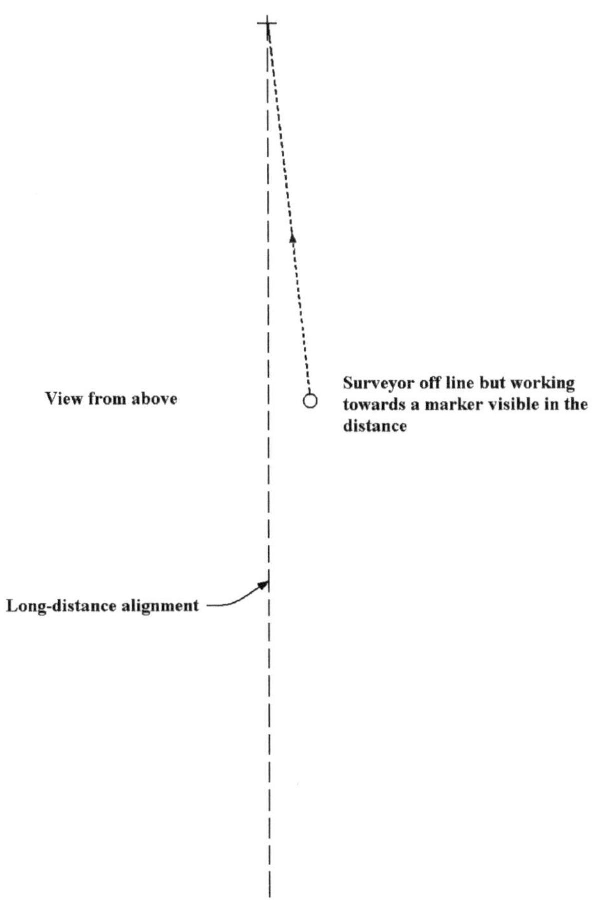

Figure 37 A surveyor, having strayed slightly off a long-distance alignment but working towards a marker visible in the distance, will tend to converge upon the long-distance alignment

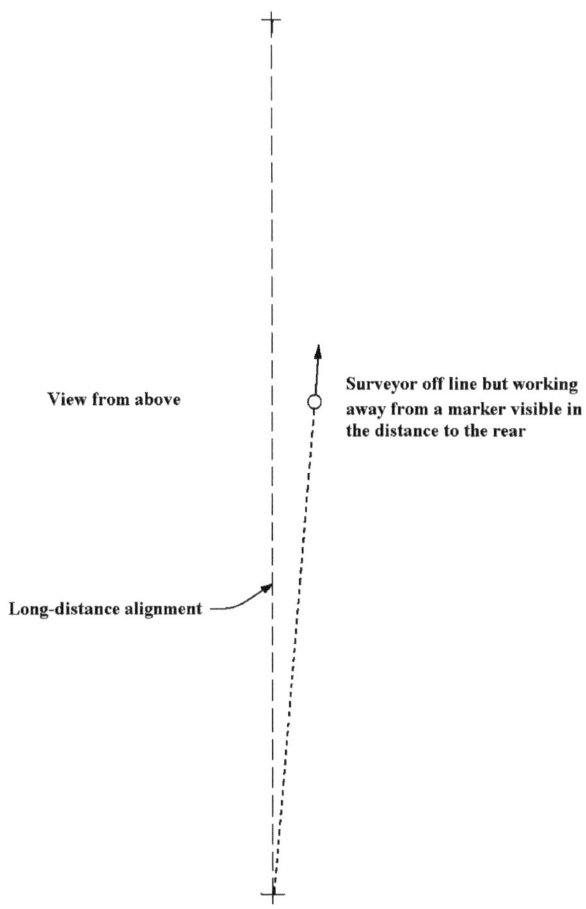

Figure 38 A surveyor, having strayed slightly off a long-distance alignment but working away from a marker visible to the rear, will tend to diverge from the long-distance alignment

Once the guiding alignment had been marked out on the ground, there would follow the task of planning and engineering a practical course for the road along the line. As already indicated, this would have called for deviations to be made from the alignment in some places in order to circumvent obstacles or to cross rivers at suitable locations. If time should have been pressing, it can be imagined that the temptation would arise to telescope these two stages of (a) setting-out and (b) planning, with the result that attempts would have been made to mark out the intervening alignment and lay out the course of the road at the same time.

If so, it can be seen that this would cause difficulties when returning the course of the road from a deviation and attempting to resume the guiding alignment, especially if neither of the distant high points should have been visible at that point, or possibly only the one to the rear. The upshot would have been that in some places the course of the road may not have followed the original alignment exactly, but run a few yards or metres to either side of it. In practical terms this would not have been significant. Indeed, in some places it may have allowed a slightly easier choice of land, and, once the road had been completed, it might not have been noticeable or regarded as important by the Roman high command. Instances of such small divergences from overall alignments can be observed in all three of the Roman roads described in Sections 7 to 9. On the other hand, where Roman Dere Street appears to follow its underlying long-distance alignments, it appears to do so exactly, for the most part. This may indicate, as suggested above, that those alignments underlying the course of Dere Street had already been quite closely marked out on the ground beforehand.

Finally, there is one more reason why such small divergences from the original long-distance alignments may have occurred. As already indicated, when the original exploratory survey lines were being set out, it is envisaged that the Roman surveyors would have been sighting from high point to high point, sometimes over considerable distances. When the road planners were setting out the alignments of their roads, however, it appears that they would naturally tend to change direction not at the highest points of hills but at brows from which they could get views not only into the far distance but also of the foreground and middle distance as well. Where changes of alignment were required, this would cause them to set out alignments slightly at variance to the original survey line, as illustrated in Figure 39. An instance of this may have occurred with the Roman road described in Section 9.

Use of guiding alignments set out by the road planners themselves

If the long-distance alignments had been set out by the road planners themselves, simply as guides for the overall directions that their roads should take, then all of the factors described above – with regard to the re-use of exploratory survey lines – would have applied in the same way. If there had been no instructions from higher authorities to utilise particular survey lines, however, then the road planners may have felt less compunction to follow their own long-distance alignments so exactly. For them, the alignments would have served their purpose adequately simply by setting out the general directions that they should take. Being less scrupulous about following their long-distance alignments precisely would have allowed the road planners more freedom to adopt easier ground along the way, as well as being able to take less trouble with any intermediate marking-out of the alignments. This would have enabled their teams to get the road-building job done more quickly. Examples of this may have occurred in places with all three of the roads to be described in Sections 7 to 9.

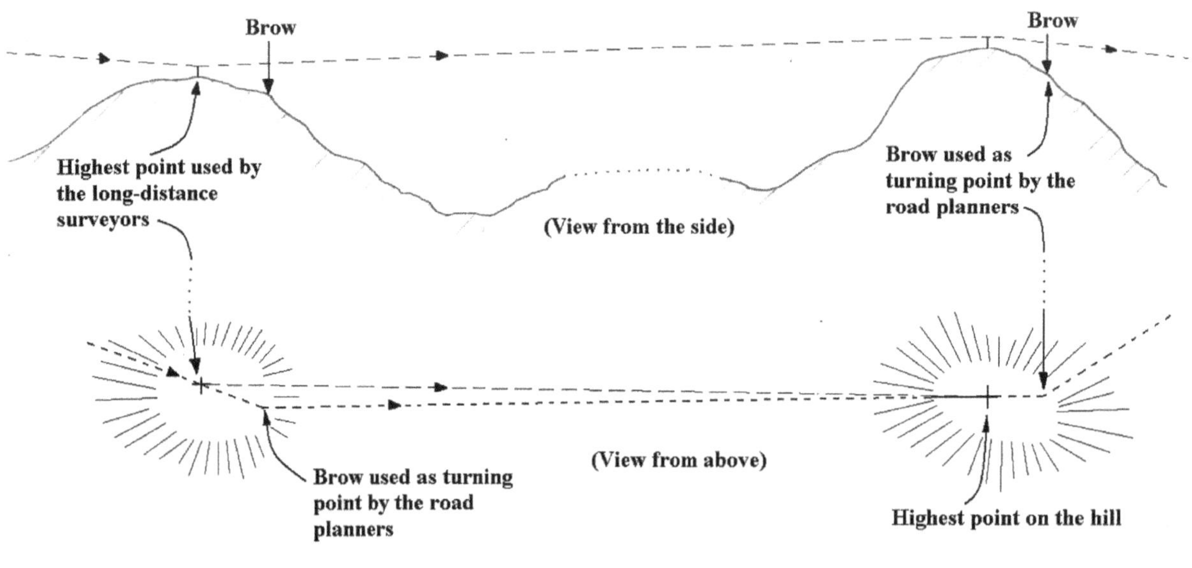

Figure 39 Potential for slightly different alignments to arise between survey lines and roads because exploratory surveyors would work from hilltop to hilltop whereas road planners would tend to work from brow to brow

Section 7: Example of the surveying and planning of the Western Main Road from Manchester through Low Borrow Bridge to Kirkby Thore, and then of the Maiden Way onward to Carvoran

The Western Main Roman Road ran for some 87 miles (139 km) from Manchester past Ribchester and Low Borrow Bridge to – it is now understood – the Roman fort of Kirkby Thore. Previously, having passed through the Lune Gorge by Low Borrow Bridge, this road was thought to have run northwards to the Roman fort at Brougham, near Penrith. However, recent fieldwork using LiDAR by Hugh Toller and colleagues has established that, north of the Lune Gorge, the main road bore off more to the north-east, towards the centre of the country.[63] If there had been a Roman road to Brougham, it now appears likely to have been a branch from the main road. See Figure 40 and also Figure 53.

The Western Main Road (which is the name that will be used for it throughout the rest of this section) was one of the two north-south trunk Roman roads in northern England. To a large extent, it ran to the west of the Pennine Hills, whereas the other trunk Roman road, Dere Street, stayed to the east of the Pennine chain. As it happens, however, at Kirkby Thore, the general line of the Western Main Road was continued for another 26½ miles (42.4 km) by another Roman road, called the Maiden Way, which did cross the Pennine chain to service the Roman forts at Whitley Castle and Carvoran.

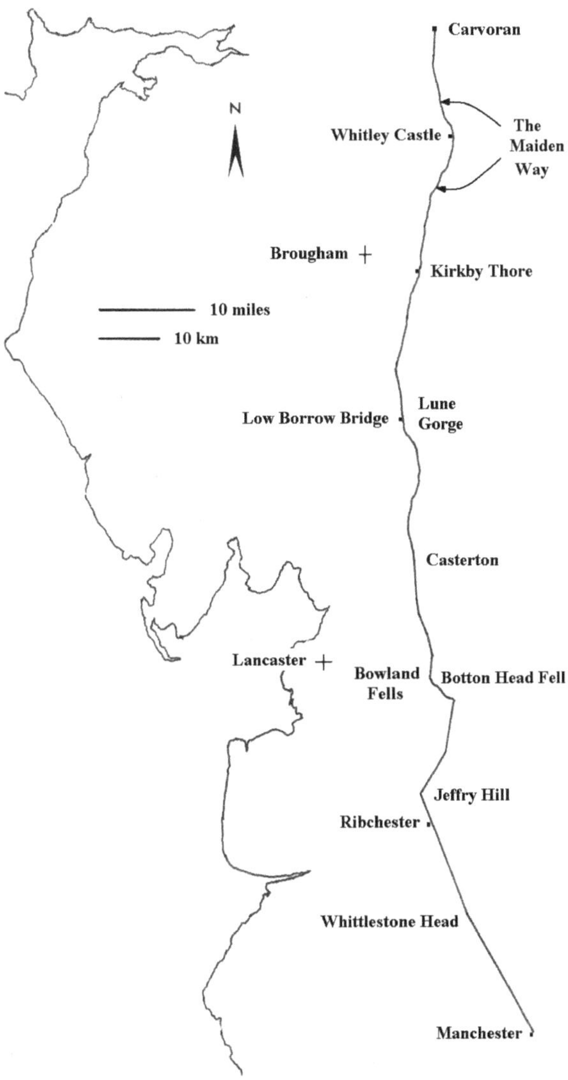

Figure 40 Overall course of the Western Main Roman Road from Manchester to Kirkby Thore and then the Maiden Way onward to Carvoran

[63] Toller, 2014, forthcoming

Unlike Dere Street, the Western Main Road has not acquired a post-Roman name. Unlike Dere Street, too, relatively little of its course has been adopted by modern main roads. Nevertheless it was built to main road standards. Immediately to both the south and the north of the Lune Gorge, for instance, its ditches flanking the road appear to be more than 30 feet (9 m) apart.

To some extent the Western Main Road seems to run across the grain of the countryside, in directions which are different from the ways in which communications have developed since the Roman period. This included making an exposed high-level traverse of the Bowland Fells over Botton Head Fell which, it can be imagined, would have been abandoned quite rapidly as a through route once it had lost its importance to the Roman military. However, it is believed that the choice of this unusual traverse and other features of the course of this road can be explained by the long-distance planning process. After extensive modelling of various possibilities it is believed that the following is the likeliest combination of processes by which the course of this road was set out.

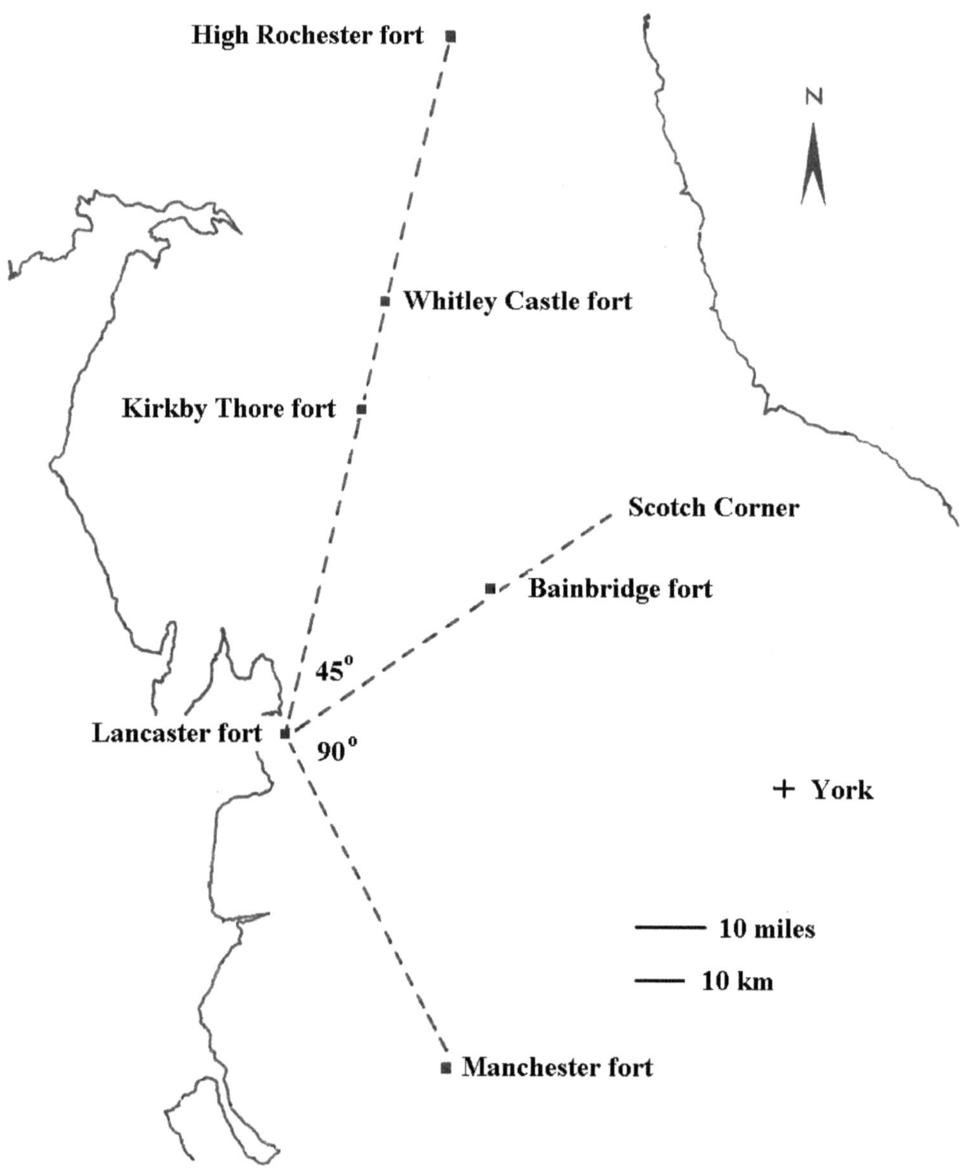

Figure 41 Exploratory survey lines radiating from Lancaster

Long-distance planning framework

By the time that the course of the Western Main Road came to be planned, it can be assumed that the three long-distance alignments which Rob Entwistle had identified as radiating from Lancaster were already in existence.[64] See Figure 41. Before examining how these alignments appear to have been used, it is worth considering why, in the opinion of the author (and also of Rob Entwistle), these alignments are unlikely to be imaginary, and what their purpose or purposes may have been.

According to the account in Margary,[65] the Western Main Road led off from the fort at Manchester in a north-easterly direction, but then, within less than a mile (approximately 1.2 km), it began to swing north-westward near the site of the cathedral. This turn appeared to curve on to an alignment from Lancaster, some 44 miles (71 km) distant. As already indicated, Lancaster is completely invisible from Manchester, any view between the two sites being blocked by the hills above Darwen and outliers of the Bowland Fells near Oakenclough. Further, having turned on to this alignment, the Roman road then followed it

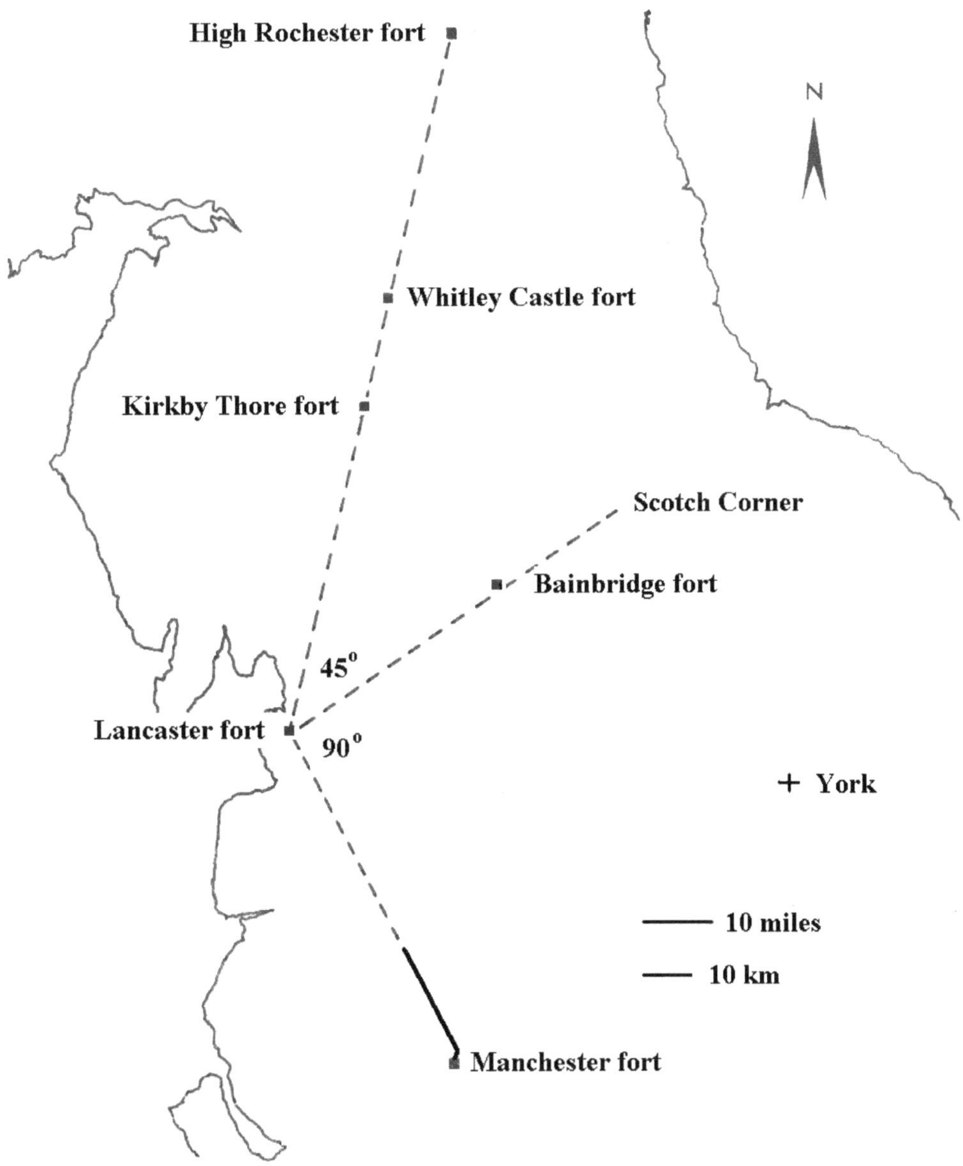

Figure 42 The Western Main Roman Road begins by leaving the fort at Manchester and then turning to run for some 14 miles directly towards Lancaster

[64] This is not necessarily meant to imply that a Roman fort would have been in existence at Lancaster at the time when the alignments were set out, although, as will be discussed shortly, it would have been obvious that the site would come to be occupied by a fort in due course.

[65] Margary, 1973, 370

for almost 14 miles (22.5.km) towards Lancaster, with only relatively minor divergences along the way. See Figure 42. It is believed that these divergences can be explained by responses to the intervening topography, which will be examined in detail later in this section. The possibility thus appears to be negligible that the Roman road, having worked its way northwards from Manchester, should by coincidence appear to have turned exactly upon an alignment towards Lancaster and then followed the alignment for nearly 14 miles (22.5 km). There can therefore be little doubt, it is considered, that the alignment from Lancaster to Broughton (just north of Manchester) did exist in Roman times.

The other two alignments observed to be radiating from Lancaster might more easily be considered to be coincidences. As already noted, although four Roman forts came to be located along the northerly alignment, two of the forts – at Kirkby Thore and Whitley Castle – do not lie exactly upon its line. Similarly, the central alignment passes at some distance from the Roman fort at Bainbridge in Wensleydale. However, these two alignments are at almost exactly 45° to each other, and the central alignment is at an almost exact right angle to the alignment towards Manchester. These are the kind of set angles which could easily have been reproduced by the Roman authorities both in the field and at their headquarters. Taken together with the foregoing evidence for the existence of the southern alignment, therefore, there appear to be good grounds for considering that all of these three lines are not imaginary but that they did constitute a network of alignments that radiated from where the Roman fort and later the medieval castle came to be sited at Lancaster. Later in this section, further evidence that these lines existed in Roman times will be described.

Why might these alignments have been set out? As already discussed in Section 5, since the southern alignment does not run exactly between the Roman sites at Lancaster and Manchester, it appears to be unlikely that a precise bearing between the two locations had been established by an earlier survey. However, there is just the possibility that there had indeed been an earlier survey, and that the initial course that the Western Main Road had taken to the north-east of the fort at Manchester had been adopted in order to circumvent some awkward bends in the River Irwell before making the turn to head directly for Lancaster. This possibility would have been quite practical in terms of road planning, but such a survey from Manchester to Lancaster would need to have covered a lot of ground, and would probably have taken a number of years to complete. Such a timescale may be considered unrealistic at a time of conquest, which is judged to have been the most likely period in which the survey lines may have been set out. On balance, therefore, it appears to be more likely that the three lines radiating from Lancaster were the result of survey lines which were set out to undertake rapid assessments of the countryside to the south, east, and north of that point.

The site which came to be occupied by the Roman fort at Lancaster is a natural strong point. Easily accessible from the sea, it resides on a hill overlooking the harbour and a crossing-place of the River Lune. Moreover, the location is a strategic one, in that, because of the elevation of the surrounding landscape, virtually all of the traffic and armies travelling between the south and north on the western side of the country would have been obliged to pass by its foot. Hence it would have been a natural site for the Romans to have seized upon, and from which Roman survey parties could have been sent out to explore the hinterland.

The next question is: what may have caused the Roman surveyors to choose the directions in which to set out the alignments for their surveys? From the top of the hill at Lancaster there are no visible features standing out in the landscape today which might have prompted the Roman surveyors to have set out either the northern or the southern of the three radiating alignments in their particular directions. For the middle alignment, too, the view eastwards would have been blocked by a ridge of high ground immediately to the east of Lancaster, which is known as The Ridge.

On the other side of The Ridge, however, overlooking a sharp double bend of the river – known locally as the Crook o' Lune – the River Lune upstream of the bend can appear to be flowing down directly from between the two prominent Pennine Hills to be seen in the distance, which are known nowadays as Whernside and Ingleborough. This might therefore have seemed to offer to the Roman surveyors the most inviting prospect for an alignment along which to penetrate east-northeastwards into the interior of the territory from Lancaster. See Figure 43. It would not have been difficult to have set out such an alignment to run over The Ridge from a starting-point at Lancaster. Undoubtedly, it must be admitted that this model is a highly speculative one, but it does fit in with what the author has learned about Roman surveying practices in his earlier studies of Dere Street, Hadrian's Wall, and the Antonine Wall in Scotland.

If this prospect had indeed attracted the attention of the Roman surveyors, however, they were due to be disappointed. Only 5 miles (8 km) upstream, to the north of Hornby, the course of the River Lune starts to turn more northwards, and by a further 7 miles (11 km) upstream, not far from Casterton near Kirkby Lonsdale, the river is running almost due north-south. See Figure 44. Meanwhile, the projected survey line, if continued on its original alignment to the south of Hornby, would have run onto higher ground at Ribblehead and Cam Fell – something of a dead end in terms of penetrating into the interior of the country. Nevertheless, if extended further the line would then have crossed over further tracts of moorland before descending from the Pennines at or near Scotch Corner, where it may have been able to link up with the Roman planning network on the eastern side of the Pennine Hills.

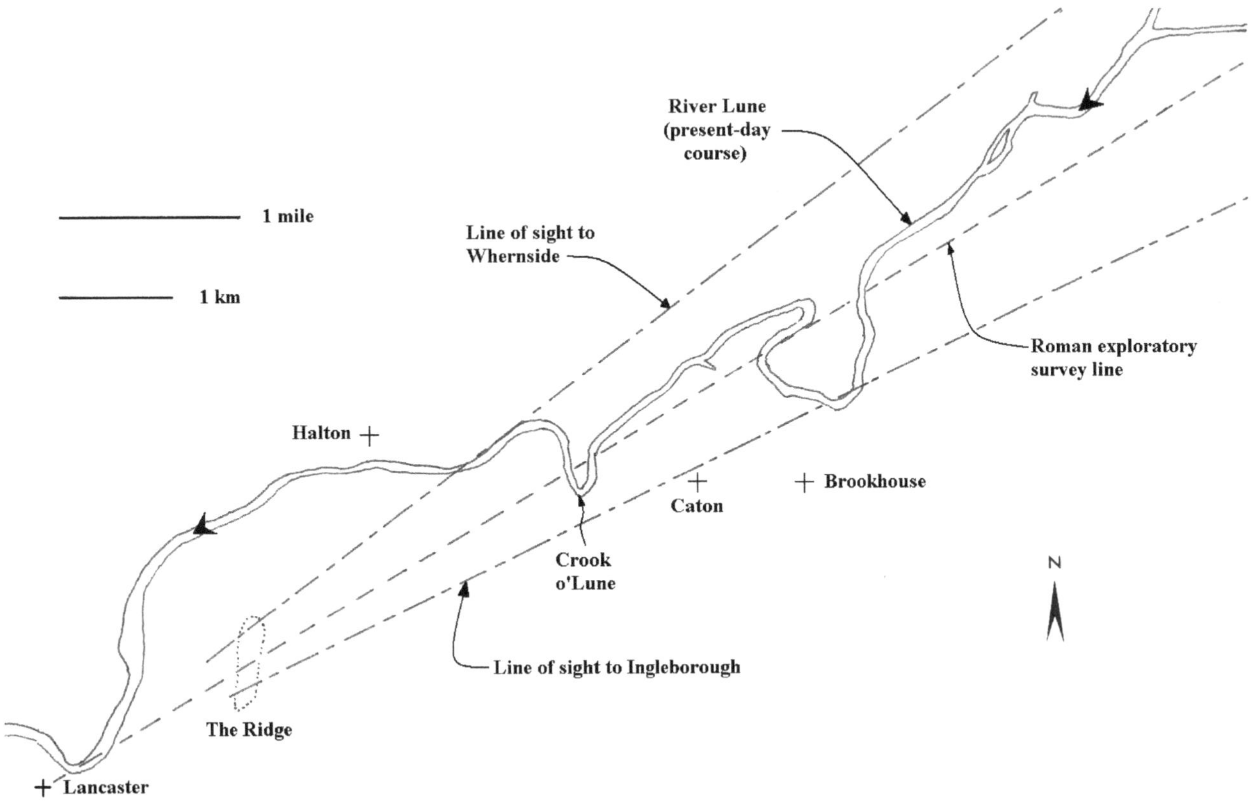

Figure 43 Possible explanation for the bearing of the exploratory Roman survey line running east-northeastwards from Lancaster

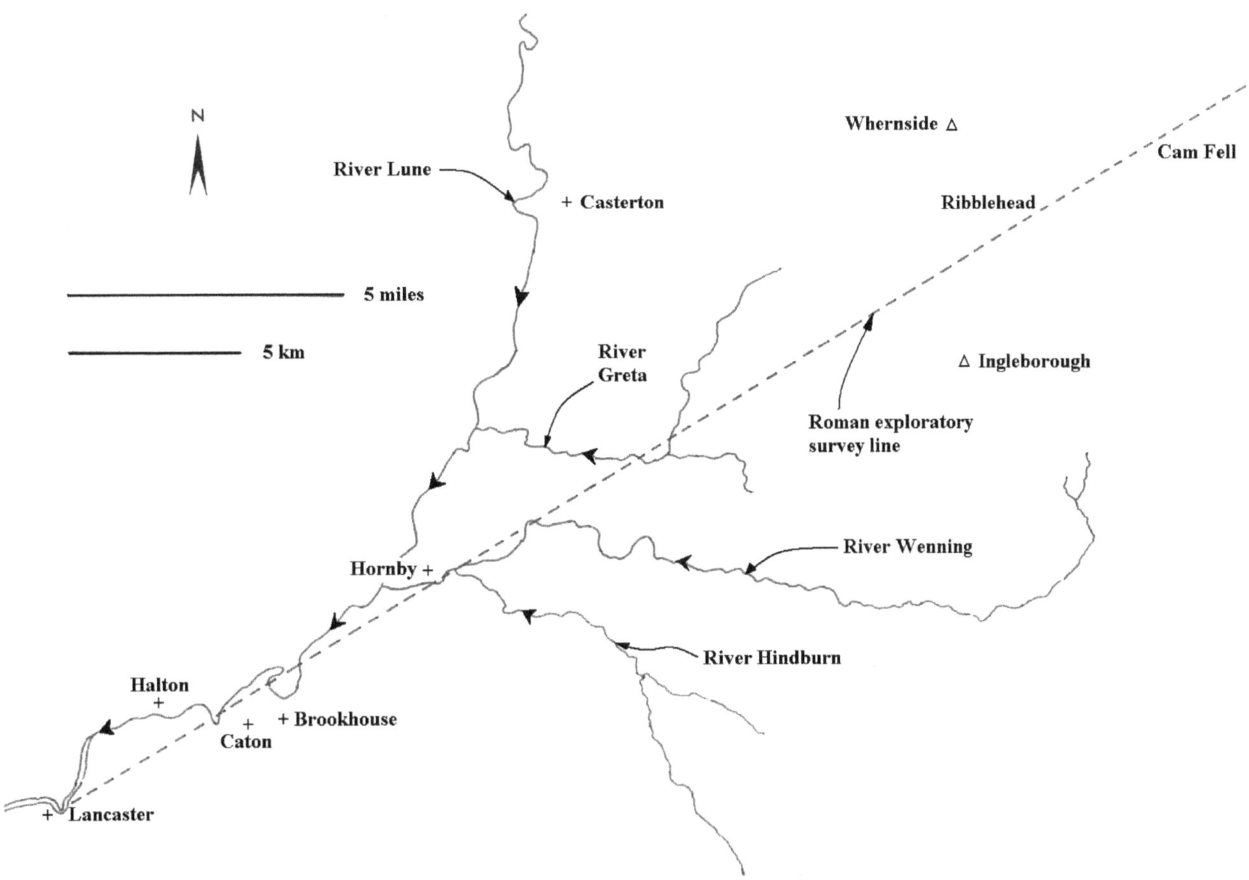

Figure 44 Course of the River Lune turning northwards away from the exploratory Roman survey line once upstream of Hornby

It will be left to Rob Entwistle to discuss the perceived Roman planning arrangements around Scotch Corner. To be left to Rob as well is the possibility that the alignment from Lancaster may have been used to help determine the location of the Roman fort at Bainbridge in Wensleydale, and also the course of the Roman road which is reputed to run to that fort from near Ingleton. Attention in this monograph can now be turned to considering how the Western Main Road may have been set out, seemingly using the information received from the projected survey lines radiating from Lancaster.

From Figure 1 it can be seen that both the Western Main Road and Dere Street keep well inland from the coast. It may be suspected that this was the result of a deliberate Roman policy.[66] Certainly, the supposition that this had indeed been Roman policy, together with the strictness with which it appears to have been enforced, are crucial to the interpretation of the planning of the Western Main Road which follows.

The effect of this presumed Roman policy can be observed straight away. As already noted, the Western Main Road, having started out north-eastwards from Manchester, then turned and headed directly towards Lancaster for several miles. Nevertheless, if the Roman policy had been that the road should stay well away from the coast, then it would have been apparent to the planners of the road that its course should turn northwards at some point, well before it reached Lancaster.

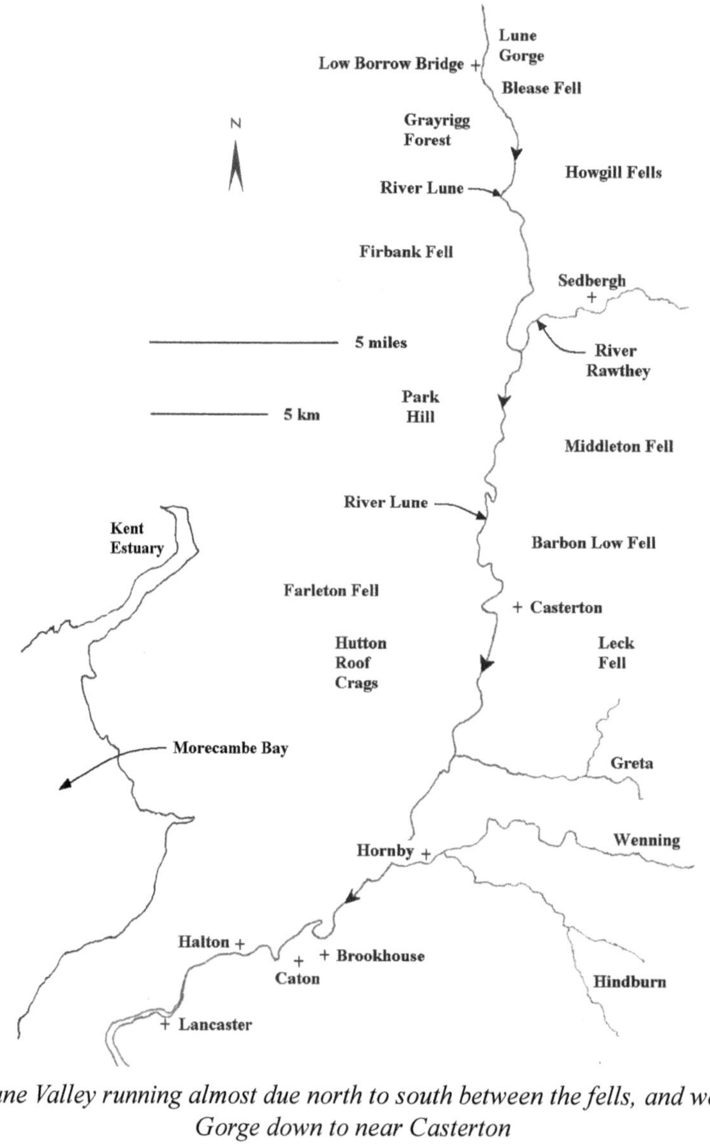

Figure 45 The upper Lune Valley running almost due north to south between the fells, and well inland, from the Lune Gorge down to near Casterton

[66] David Breeze, *pers. comm.*, has suggested that the Romans may have kept their roads well inland from the coast in order to avoid having to cross many streams and rivers near their widest points. An alternative possibility is that the roads might have been sited inland near the heads of navigation of the major rivers, so facilitating the provision of supplies by water to the military stations along the roads To the author, however, and to judge by the degree to which the policy seems to have been applied, it is felt that the military reasoning may have had a more strategic basis than this.

Figure 46 View southwards down the Lune Valley from Blease Fell, at the southern entrance to the Lune Gorge. Photograph taken in 1964-5

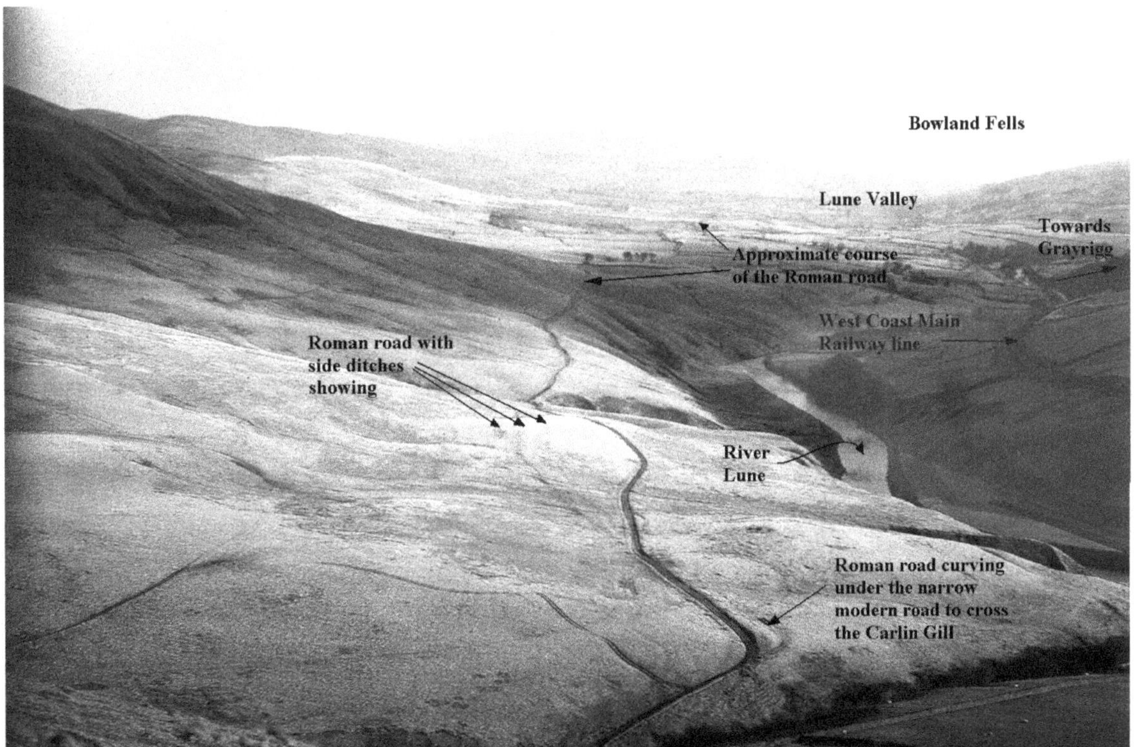

Figure 47 Annotated version of Figure 46, showing the Western Main Roman Road running down the valley of the River Lune whereas the West Coast Main Railway line, having passed through the Lune Gorge, then exits the Lune Valley by a side valley heading towards Grayrigg and then Oxenholme, near Kendal

At this juncture, the next stage of the long-distance planning process may well have shifted well to the north of Lancaster. If the surveyors following the initial alignment of the River Lune near Lancaster may have been disappointed by the course that the river was found to be taking further north, the behaviour of the river upstream of Casterton may have been a real boon to the planners of the Western Main Road. See Figure 45. Running almost north-south, and suitably inland from the coast, the upper Lune Valley could have seemed an almost perfect corridor for the road to take.

Moreover, at its northern end, the valley runs into the Lune Gorge, which is the only all-weather pass between where the Pennine Hills to the east come close to touching the Lakeland Fells to the west. This pass is now used by both the West Coast Main Railway line between London and Glasgow, and the M6 Motorway. The only significant main road which attempted a different route to the north in this area was the A6 road between Kendal and Penrith. Older readers may remember that this road could almost be guaranteed to become blocked by snow every winter, with consequent disruption to all north-south trunk traffic on the western side of England before the M6 was built.

Figure 46 shows the view looking southwards down the Lune Valley from the mouth of the Lune Gorge. It is not an aerial photograph. The picture was taken from about two-thirds way up Blease Fell, which stands at the southern entrance to the Lune Gorge, and it presents a view down the Lune Valley which a Roman surveyor could have obtained as readily as did the author.

As can be seen on the view, the West Coast Main Railway line, having passed southwards through the Lune Gorge then exits the Lune Valley by a side valley towards Grayrigg. This was in order for the line to pass close by Kendal, at Oxenholme, before travelling down the coastal strip via Carnforth to Lancaster. The photograph was taken in the 1960s, before the M6 Motorway was built, but that road too exits the Lune Valley via the same side valley, but then takes a steeper but more direct route via Killington Reservoir past Carnforth to Lancaster. In contrast, the Roman road continues directly down the Lune Valley, as shown on Figure 47.

The problem facing the planners of the Roman road would have been how to get it out of the Lune corridor at its southern end. This was where the Rivers Greta, Wenning, and Hindburn come down from the Pennine Hills to the east to join the Lune and then run westwards with it to Lancaster. If the Roman road planners had strict instructions to stay inland from the coast, they would have had to find a way out of the Lune Valley keeping more or less due south. However, at this point in the landscape the Bowland Fells appear to be a solid wall blocking the end of the north-south corridor, with no obvious gap in them which might suggest a way through.

The solution which seems to have been adopted by the planners of the Roman road was to proceed down the Lune Valley until the river began its turn to the west, and then to set out an exploratory planning line almost due south that ran straight up and over the Bowland Fells. This line appears to have been set out from a point to the south of Casterton, where the ground begins to fall to lower-lying land across which the Leck Beck and then the Greta, Wenning, and Hindburn Rivers run before joining the Lune. See Figure 48.

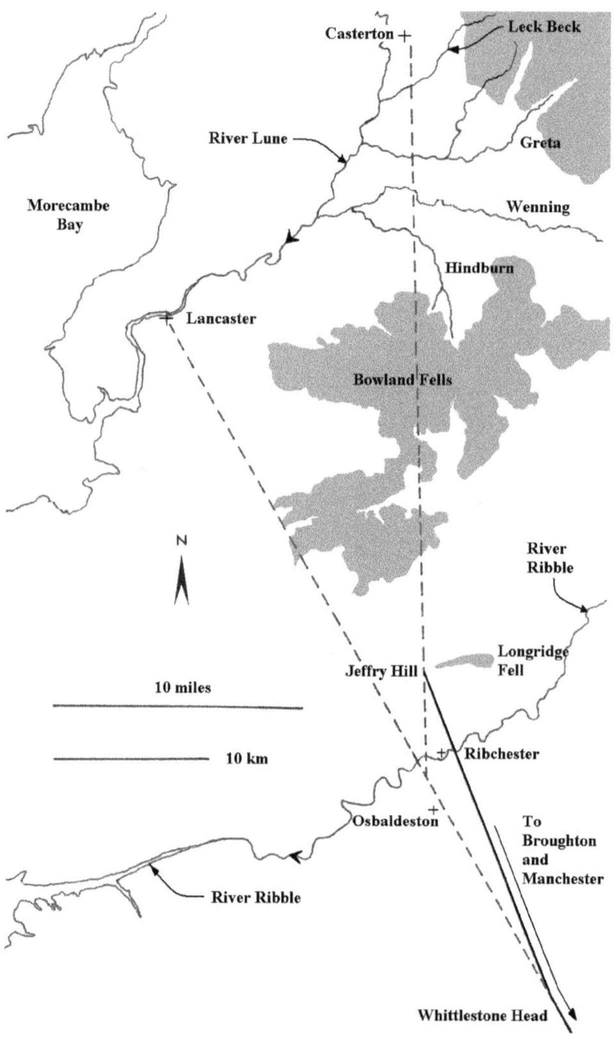

Figure 48 Long-distance planning line laid out over the Bowland Fells almost due south from a point near Casterton. The intersection of this line with Longridge Fell at Jeffry Hill appears to have enabled the Western Main Roman Road to take a short-cut from near Whittlestone Head to Jeffry Hill

It may be that the intention had simply been to continue this exploratory alignment almost due south until it should meet the line of the road coming up from Manchester, along its alignment to Lancaster. If so, these two alignments would have met somewhat to the south of the River Ribble, about a mile or so (1 to 2 km) north of the village of Osbaldeston. See Figure 48 again. However, after having crossed the Bowland Fells and then the valley of the River Loud, the alignment from near Casterton happened to run up against and on to a prominent east-west ridge, known as Longridge Fell, at a place called Jeffry Hill. Longridge Fell can be seen for miles around. In particular, it can be seen from where the alignment coming up from Manchester to Lancaster crosses high ground near Whittlestone Head, to the south-east of Darwen. It appears that this enabled the planners of the Roman road to set out a short-cut from this high point to Jeffry Hill, rather than continuing to Osbaldeston.

It was Rob Entwistle who pointed out these alignments to the author. In particular, he observed that the first 3½ miles (5.6 km) of the Western Main Road to the south of Casterton were, as built, exactly aligned upon Jeffry Hill. Jeffry Hill is completely invisible from Casterton, standing some 24 miles (38.9 km) distant on the south side of the Bowland Fells. It is considered, therefore, that there can be little doubt that this alignment is not imaginary.

In terms of planning there appears to be little doubt, too, that the alignment from near Casterton was set out from north to south. There is no physical feature at Jeffry Hill which might have prompted Roman planners to select it as a sighting-point. Nor are there any features on the Bowland Fells facing Jeffry Hill which might have prompted a surveyor to choose a line in any particular direction there. Moreover, if the Roman planners should, arbitrarily, have chosen to set out a line almost due north from Jeffry Hill, they would have been extremely fortunate to have landed in such a well-positioned location for the continuation of the road up the Lune Valley. Finally, the starting-point of the alignment near Casterton is on a south-facing slope, with little view to the north. All of this evidence supports the probability that this alignment had been set out from north to south.

In contrast, the direction of the planning of the alignment from Jeffry Hill to near Whittlestone Head is less clear. When observed from Jeffry Hill, the point at which the change of alignment was made near Whittlestone Head is not the highest position on the skyline. Neither does the point possess a physical feature which could have made it a target for any sighting from Jeffry Hill. Indeed, unless the position of the turning-point should have been indicated by a light at night, it would have been difficult to have set out a line to it from Jeffry Hill.

Alternatively, once the location at Jeffry Hill had been established, as it appears, by the line from Casterton, this could have been used as a target for an alignment from near Whittlestone Head. However, when viewed from near Whittlestone Head, Longridge Fell is not the highest point on the skyline. The Bowland Fells loom higher behind it. Thus, when seen from near Whittlestone Head, the featureless location of Jeffry Hill, some 13 miles (21 km) away, would almost certainly have made an invisible target unless it should have been indicated by a light at night. Either way, these options appear to offer another indication that the Romans undertook some of their long-distance surveying using lights at night.

With the setting-out of the alignment between Jeffry Hill and the high ground near Whittlestone Head, the framework of long-distance alignments for the Western Main Road would have been completed between Manchester and the fort at Low Borrow Bridge, which lies in the centre of the Lune Gorge. See Figure 49. However, it can be imagined that the process of laying out the planning alignment from Casterton to Jeffry Hill would have left the planners of the road with no doubts about the unsuitability of such a direct line across the Bowland Fells for an all-weather road. At least 12 miles (19 km) of high-level walking would have been involved, with no shelter from the elements, especially in the winter.

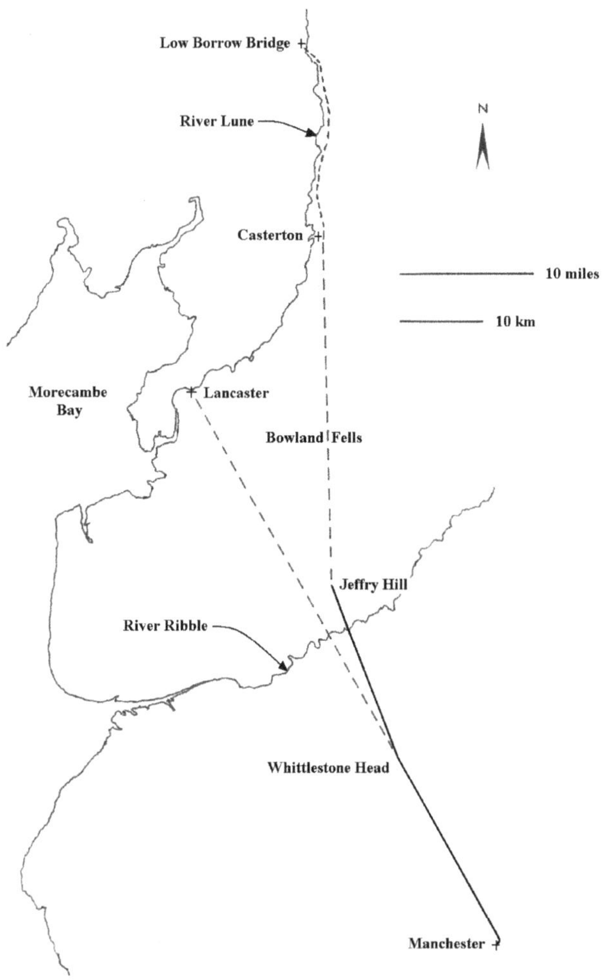

Figure 49 Long-distance framework for the Western Main Roman Road between Manchester and Low Borrow Bridge in the Lune Gorge

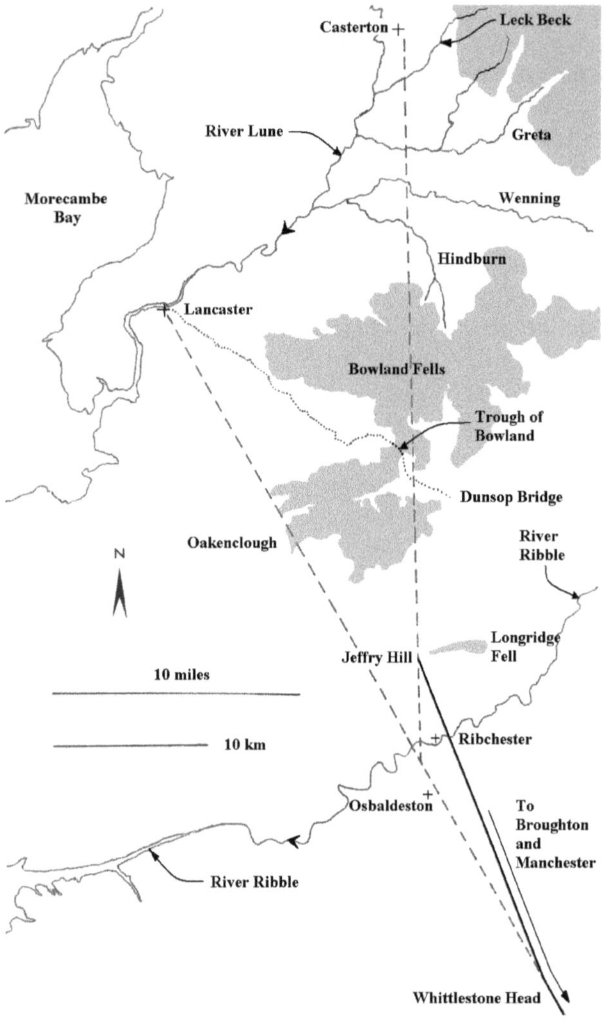

Figure 50 Shown dotted, the road through the Trough of Bowland offers the only real pass through the Bowland Fells

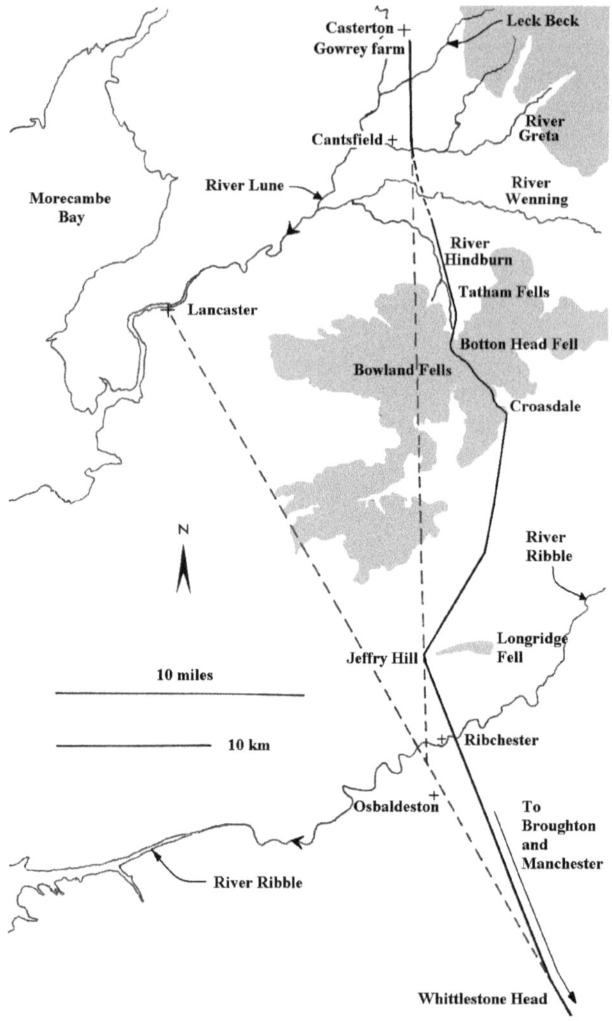

Figure 51 Deviation of the Western Main Roman Road up Croasdale, over Botton Head Fell, and then down past the Tatham Fells to rejoin the long-distance planning line near Cantsfield

Standing at Jeffry Hill and looking down the steep northern slope of Longridge Fell, the low-lying fields below can be seen stretching out to the north and north-west all the way to Morecambe Bay. Lancaster itself is hidden by an outlier of the Bowland Fells at Oakenclough, but it is very obvious where the sensible course of the Western Main Road should run – across the easy-going low-lying land between the Bowland Fells and the sea.

However, if the instructions had been to keep the road inland from the coast, then the planners of the road would have had to find practical route through the fells instead. As it happens, one pass through the Bowland Fells does exist. Known as the road through the Trough of Bowland, it enters the hills near Dunsop Bridge but, it then, unfortunately, terminates at Lancaster. See Figure 50. Again, it is assumed that this would have been unacceptable to the Roman military authorities. Therefore, no doubt after much searching, the planners of the Western Main Road seem eventually to have found a route across the Bowland Fells by going up Croasdale and then over the top on to Botton Head Fell before descending via the Tatham Fells to the Lune Valley below. See Figure 51. At its highest point this route would still have been very exposed to inclement weather, but at least the exposure would have been more brief than that which would have been incurred by following the long-distance planning line.

Thus, as can be seen in Figure 51, the planners of the Western Main Road appear to have set out a deviation from Jeffry Hill to run across to and then up Croasdale, then over Botton Head Fell and down by the Tatham Fells in order to rejoin the planning line near Cantsfield, just west of Burton in Lonsdale. From there the planning line was followed exactly all the way to just north of Gowrey farm, south of Casterton. In addition, the way in which the Western Main Road, as built, curves gradually on to the planning line near Cantsfield is reminiscent of the way in which Dere Street, as built, curves on to its planning line from Tadcaster at Dishforth. Both of these features are felt to indicate that the planning lines had been in existence before the courses of the roads, as they were subsequently built, had been set out on the ground.

Northwards up the Lune Valley from near Casterton, no long-distance alignment would have been needed for the Roman road. With the Pennine Hills encroaching from the

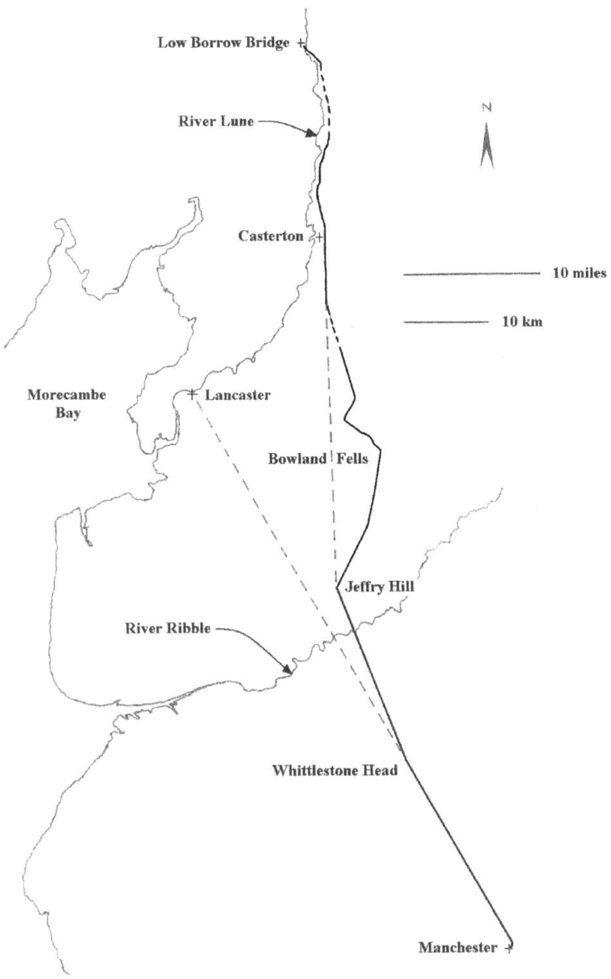

Figure 52 No long-distance alignment required for the Western Main Road between Casterton and Low Borrow Bridge: the encroaching fells and the river largely dictate the course of the Roman road here

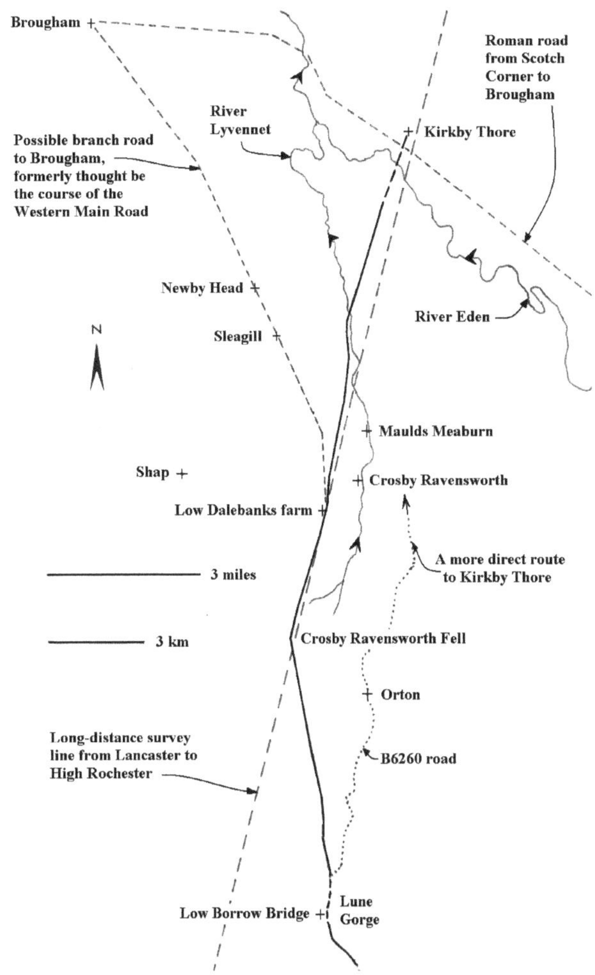

Figure 53 Course of the Western Main Road north of the Lune Gorge, showing the line now established by LiDAR as heading for Kirkby Thore rather than Brougham as thought previously

east and the River Lune flowing on the west, the course of the road would have been dictated by the increasingly narrow stretch of land between the hills and the river, and, no doubt, the areas of marsh scattered along it. See Figure 52.

North of the Lune Gorge, the Western Main Road does not set off up towards Shap, as do both the West Coast Main Railway line and the M6 Motorway. Instead, the Roman road noticeably keeps more to the north, seemingly so as to fall in, on Crosby Ravensworth Fell, with the northerly survey alignment running from Lancaster to the forts at Kirkby Thore, Whitley Castle, and High Rochester. See Figure 53.

Until the recent fieldwork by Hugh Toller and colleagues, it had been believed that after crossing Crosby Ravensworth Fell, the Western Main Road had then turned west of north again to run via Sleagill and Newby Head to the Roman fort at Brougham, just east of Penrith. However, as described by Margary,[67] this proposed course had always struck the author as a rather unconvincing one, exhibiting a quite different character from the remains of the road elsewhere. Now Hugh Toller and colleagues have been able to show, via the use of LiDAR images, that the course of the Western Main Road did continue to follow, overall, the course of the projected survey line to the fort at Kirkby Thore rather than to Brougham. The road takes a rather extended deviation from the survey alignment between Low Dalebanks farm and Kirkby Thore, but the possible reason for this is discussed in more detail in the following part of this section. The important point about the projected alignment from Lancaster is that the course of the road appears to veer on to it before heading for Kirkby Thore. If, after leaving the Lune Gorge, the objective for the road had simply been to head for Kirkby Thore by the most direct route, then, as indicated on Figure 53, a course more akin to that of the B6260 road through Orton and then a more northerly line to the east of Crosby Ravensworth and Maulds Meaburn could have been taken instead.

[67] Margary, 1973, 386

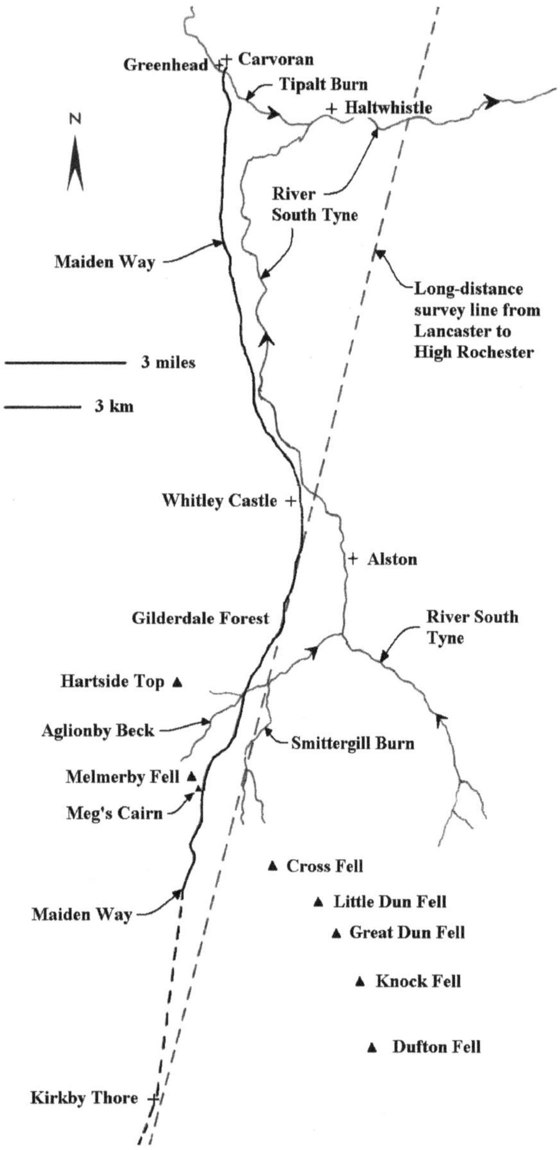

Figure 54 Course of the Maiden Way Roman road from Kirkby Thore via Whitley Castle to Carvoran

Beyond the fort at Kirkby Thore, we can now see that the course of the Western Main Road may have been continued into the valley of the River South Tyne by the Roman road known as the Maiden Way. Immediately facing the course of this road to the north-east, though, is the intimidating wall of Pennine Hills running from Knock Fell, past Great and Little Dun Fells and Cross Fell to Melmerby Fell. The projected survey line would have run on to the shoulder of Cross Fell. At just under 3000 feet (893 m), Cross Fell is reputed to be the highest hill in England outside the Lake District. As with the crossing of the Bowland Fells, this alignment would have offered a thoroughly impractical course for an all-weather road. Slightly to the north of it, the least impractical and fairly direct course across the fells passed through a slight saddle below Melmerby Fell, by Meg's Cairn, and the alignment of the Maiden Way appears to make straight for this point upon leaving the fort at Kirkby Thore. See Figure 54.

On the other side of Meg's Cairn, the course of the Roman road makes superb use of a ridge of higher ground between Aglionby Beck and Smittergill Burn before falling to cross the former just after it has been joined by the Rowgill Burn, below Meathaw Hill. The road then rises on to the open moorland known as the Gilderdale Forest, where it appears once again to veer on to the projected alignment from Lancaster and then to follow it for some 1½ miles (2.5 km) before finally deserting the line to run past the fort at Whitley Castle. Thereafter the Roman road follows the valley of the South Tyne to the Roman fort at Carvoran, which overlooks the Tipalt Burn and the village of Greenhead, to the west of Haltwhistle.

It should be noted that Carvoran is situated almost exactly half way between Carlisle on the west and Corbridge to the east, and that the Tipalt Burn and the South Tyne Valley would have formed an obvious route for the penetration of insurgent forces into England from the north between the two main Roman thrusts into Scotland. Hence the fort at Carvoran could be seen as having been set to plug this route at the downstream end whilst Whitley Castle oversaw the South Tyne's headwaters. This appears to be a more significant explanation for the existence of the fort at Whitley Castle than the one which has been proposed up to now, which is that it may have been sited mainly to protect Roman mining interests in the area.[68]

As far as has been detected, the Romans appear to have made no more use of the projected survey line to the north of Whitley Castle – except, that is, seemingly to locate the fort beside Dere Street at High Rochester. However, it has often been speculated that a Roman road would have run from Whitley Castle to Corbridge, which is where Dere Street crosses the River Tyne. If so, even though the countryside covered by the alignment is scarcely propitious for such a course, it is possible that that such a road might have made use of the line in places. Conceivably, this could make a suitable subject for a LiDAR study in the future.[69]

The foregoing completes the survey of the long-distance planning of the Western Main Road between Manchester and Carvoran. What now follows is a detailed examination of how, in places, the final course of the road, as built, may have been set out on the ground on the basis of the foregoing framework.

Examination of some details of the setting-out of the road

To begin at the southern end, excavation has shown that the Roman fort at Manchester stood at Castlefield, somewhat to the west of Deansgate railway station. The Roman road running north-eastwards from the fort is believed to have joined the main Roman road running under Deansgate at a point rather to the north of Quay Street.[70] See Figure 55.

---

[68] This view most recently affirmed by Went and Ainsworth, 2013, 136-9.
[69] It is understood that Hugh Toller and Bryn Gethin have recently undertaken some work, using LiDAR, on the course if a possible Roman road between Whitley Castle and Corbridge, but currently no details are available.
[70] Bidwell and Hodgson, 2009, 98

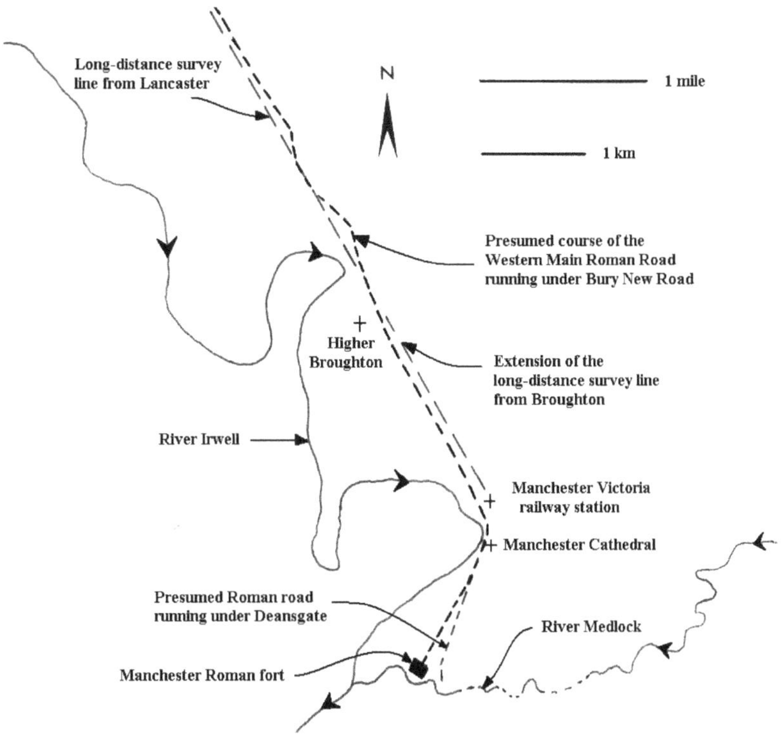

Figure 55 Presumed course of the Western Main Road in central Manchester, starting from the Roman fort at Castlefield and then turning towards the long-distance alignment from Lancaster after rounding the bend of the River Irwell beside the cathedral

At the northern end of Deansgate, starting in the vicinity of the cathedral, this main Roman road is then believed to have curved around a bend of the River Irwell, following firstly the line of Victoria Street, then Great Ducie Street, and finally coming on to the line of Bury New Road, the modern A56. According to Margary,[71] Bury New Road is a turnpike road, and it is assumed that the Western Main Roman Road may have lain underneath it for some distance here. If so, along this stretch the Roman road may have run some 150 yards (c.137 m) to the west of the perceived alignment from Lancaster. From an analysis of all the available evidence, it is considered most likely that this alignment from Lancaster, if extended southwards from Broughton, would have run close by Dutton Street towards Manchester's Victoria railway station.

With such uncertainty about the exact course of the road in this area, coupled with the fact that the surroundings are almost completely built up, it is not really practical to try to offer any topographical explanation for the apparent divergence of the road from the projected planning line here. However, to the north of Higher Broughton, as can be seen on Figure 55, the course of the Bury New Road deviates to the east, possibly to circumvent another awkward bend of the River Irwell, and then runs on to the projected long-distance alignment for a short stretch. Looking back, it can be seen that from this point southwards, the shortest practical route around the two big bends of the Irwell to the Roman fort is indeed that taken by Bury New Road, Great Ducie Street, Victoria Street, and then the branch off Deansgate down to Castlefield. Hence a deviation from the long-distance alignment to run down to the fort from this point by Higher Broughton would seem to have been eminently credible.

Turning to face northwards again, beyond its deviation above Higher Broughton, the Bury New Road then appears to run some 80 yards (c.73 m) to the east of the projected long-distance alignment before bearing away north-northeastwards and eventually losing all contact with the alignment past Whitefield. Near Radcliffe, Margary reports that the Roman road passed some 150 yards (c.137 m) west of Radcliffe Tower. This would position the Roman road still well to the east of the projected alignment, but a deviation might be expected here anyway in order to cross – finally – the course of the River Irwell at what would have been a suitable point in Roman times. See Figure 56.

Beyond this point the course of the road, as it gets into open country, becomes more certain and can be seen to be converging gradually with the projected long-distance alignment until, over the high ground at Affetside, it runs upon the alignment exactly. It is upon such high and firm ground, with good views all round, that the road can be expected to have followed a long-distance alignment more precisely, and this exact correspondence is one of the main factors which have been used for fixing the most likely line of the postulated alignment from Lancaster.

North of Affetside, however, the projected alignment runs into a landscape which is cleft by several steep-sided valleys and miniature ravines, some of which are now occupied by reservoirs. Faced with countryside like this, the planners of the Western Main Road appear to have been forced to diverge quite considerably from the alignment in places in order to adopt a practical course for the road. Even so,

[71] Margary, 1973, 370

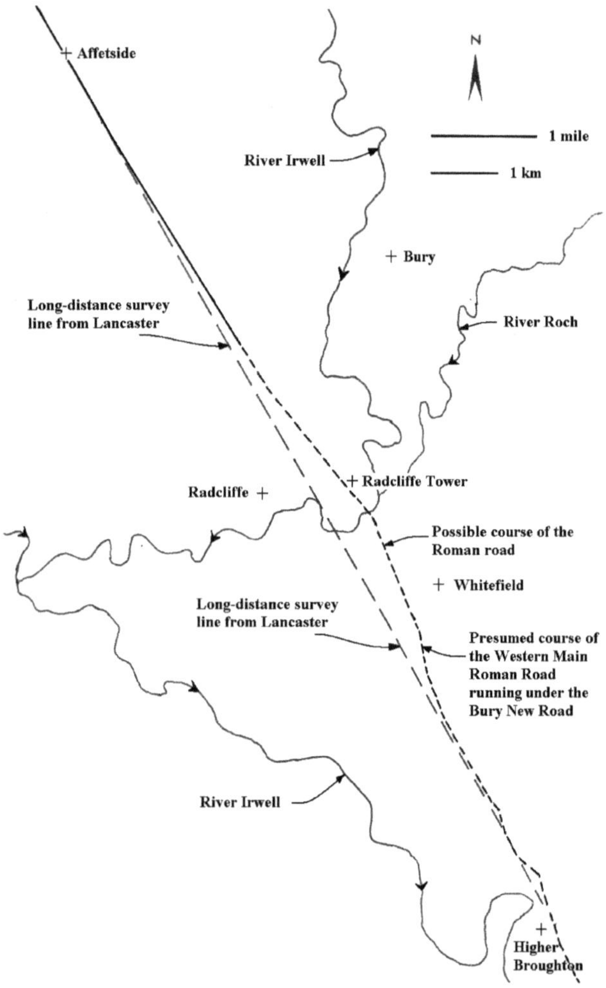

Figure 56 The Western Main Roman Road heading north-west from Manchester. The course of the road gradually becomes more certain once past Radcliffe Tower

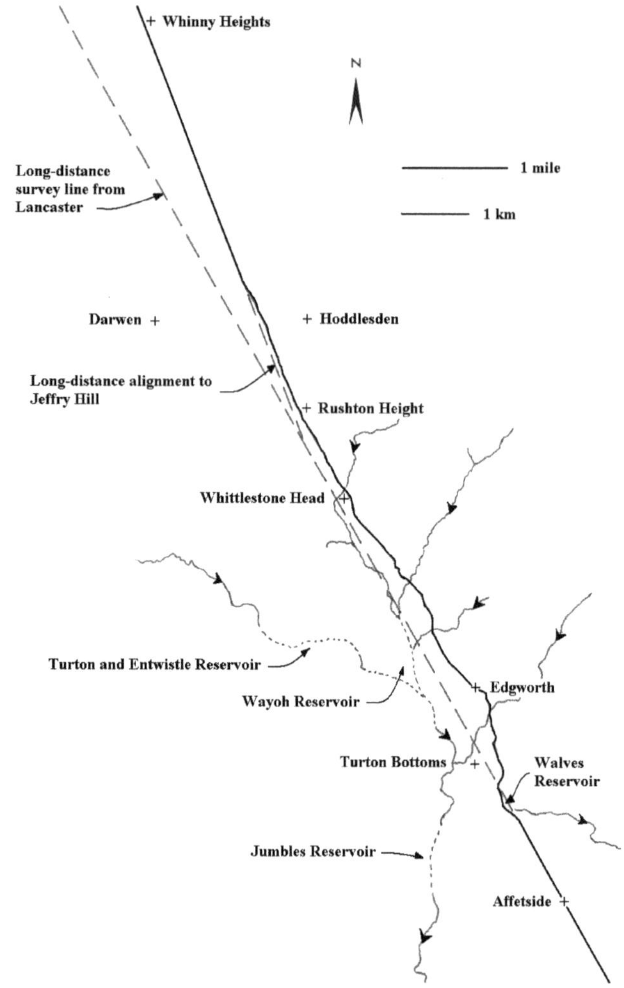

Figure 57 Course of the Western Main Road from Affetside past Whittlestone Head and on to Whinny Heights, towards Blackburn

some of the inclines and declines of the road as built past Turton Bottoms, Edgworth, and then on to Whittlestone Head are very steep, and must have provided strenuous exercise for the Roman soldiery engaged in marching upon them. Nonetheless, when examined on the ground, all of the divergences from the long-distance alignment here appear to be entirely explicable by the topography.

Perhaps anticipating the change of long-distance alignment to the north, just beyond Whittlestone Head, the divergences here were all made to the east of the alignment to Lancaster. This allowed the line of the Western Main Road to climb gradually on to the summit of a ridge of high ground near Rushton Height and then curve to fall in with the new alignment to the west of Hoddlesden. This appears to have been an eminently practical response to an instruction to follow the long-distance framework at this point. See Figure 57.

Running into Blackburn via Whinny Heights, the Roman road, as built, appears to run exactly upon the new alignment. The course of the road through Blackburn is then lost under built-up areas, but when the Roman road appears at the next high point – the Top of Ramsgreave – its line appears to lie some 80 yards (c.73 m) to the west of the long-distance alignment. See Figure 58. This is slightly surprising since the Top of Ramsgreave sits on a prominent ridge with excellent views to Blackburn and Darwen on its south side and to the River Ribble to the north. As such it would seem to have been a natural place on which to have set a marker for the alignment. Possibly, the planners setting out the line of the road on the ground may have lost sight of a marker at some previous point along the road, and strayed off the line a little. Alternatively, it is at least conceivable that an observation and signalling tower had been sited at the Top of Ramsgreave, in which case the Roman road would need to have been aligned a little to one side in order to pass by it. Whatever the explanation, the divergence appears to have been corrected by the time that the Roman road came to cross the River Ribble, since the crossing-point appears to be exactly upon the long-distance alignment.

From this crossing-point, past Cherry Yate and the cross-roads at Halls Arms Inn, the planners of the Western Main Road appear to have had no difficulty in following the long-distance alignment precisely up the southern flank

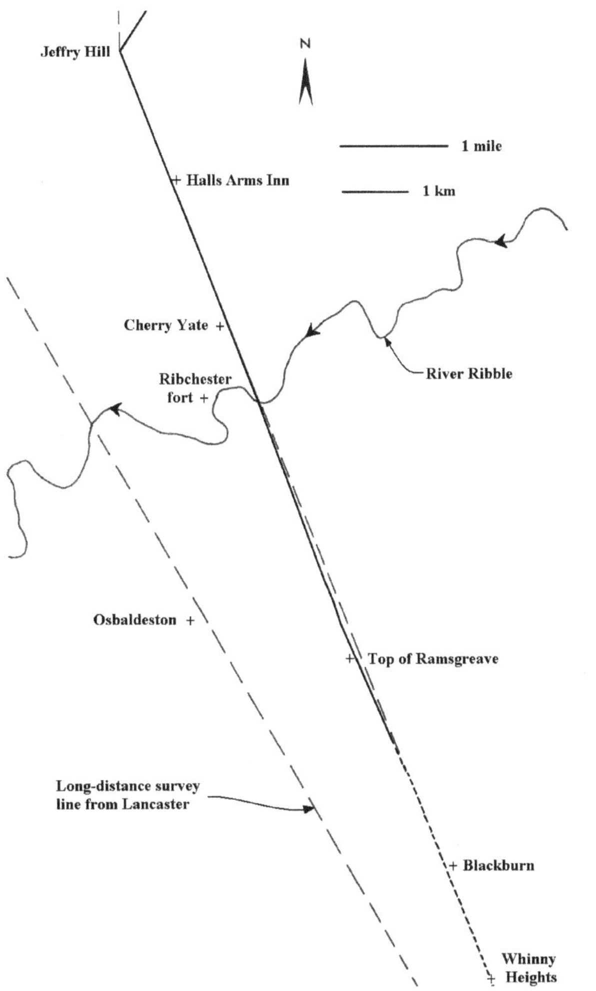

Figure 58 Course of the Western Main Road from Whinny Heights through Blackburn and over the River Ribble to the east of Ribchester, and then up to Jeffry Hill on Longridge Fell

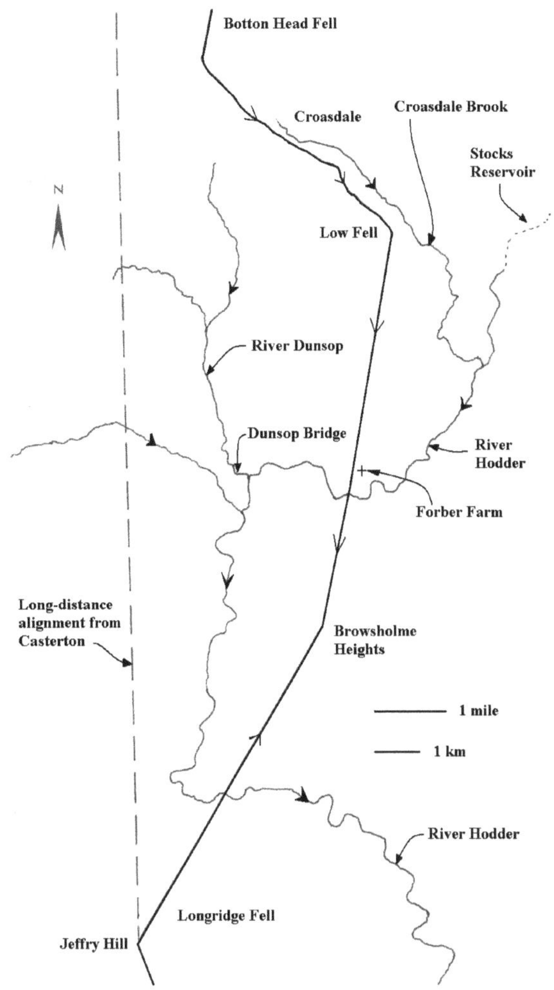

Figure 59 Course of the Western Main Road from Jeffry Hill to Croasdale and over the top to Botton Head Fell

of Longridge Fell to Jeffry Hill. It is from the north-facing brow at Jeffry Hill that the deviation was then set out to carry the Roman road over the Bowland Fells via Croasdale. The way in which the first stretches of this deviation appear to have been set out offers a fascinating example of how it is believed that the planners of the Roman road operated in practice. See Figure 59.

To begin with, even if the Romans should have wanted to continue due north from Jeffry Hill, a descent of the north face of Longridge Fell would have been impracticably steep for a road, even by Roman standards. As it happens, the entrance to Croasdale lies somewhat to the east of north of Jeffry Hill. However, the entrance to Croasdale is not visible from Jeffry Hill, the view between the two points being blocked by high ground to the west of Browsholme Heights. Hence the surveyors who initiated the course of the deviation from Jeffry Hill seem to have begun by setting out a line directly to the summit of the rise by Browsholme Heights. This line allowed the road to terrace north-eastwards down the face of Longridge Fell in a steep but practical manner, after which, as built, the Roman road followed this alignment exactly all the way to near Browsholme Heights. Since the high points of both Jeffry Hill and Browsholme Heights would have been visible from most of the stretches of road across the land in between them, this would not have been difficult.

On Browsholme Heights, a change of direction was made just to the north of the summit, implying continued planning from south to north thus far. However, to jump some 7 mile (c. 11 km) ahead, in the descent which is made down Croasdale from the road's highest point above Botton Head Fell, it is fairly clear that the Roman road had been set out from north to south. Even though the nature of Croasdale would have permitted few rigidly straight stretches to be deployed along its sides, at every significant turning point of the road down the valley, the best field of view is undoubtedly to the south. This tactic of setting out the course of a road from the highest point downwards appears to have been quite standard Roman practice. It appears to have been adopted, for instance, to the course of Dere Street from its passage across the Cheviot Hills, and to a number of places along the line of the Antonine Wall in Scotland.

As can be seen on Figure 59, the Western Main Road exits the southern end of Croasdale on a shoulder of Low Fell, quite high above the floor of the valley, and from this point Browsholme Heights would have been visible. It therefore appears that the planners of the Roman road set out a direct line from this point to Browsholme Heights. However, in so doing it would appear that they got it slightly wrong, aiming for a position some 200 yards (c.183 m) to the east of where the line from Jeffry Hill had breasted its summit by the Heights. It may be that, at a distance of nearly 5½ miles (8.8 km) from Croasdale to Browsholme Heights, the exact location of the end point of the line from Jeffry Hill could not be seen clearly, or maybe that, when viewed from Croasdale, the highest point near Browsholme Heights might have appeared to be in a different place from that when viewed from Jeffry Hill. Certainly, the line from Croasdale appears to be aligned more closely upon where the modern trigonometrical survey point was sited.

Thus a slight correction would have been called for, and this appears to have been set out from above the north bank of the River Hodder, below Forber Farm. From this point, the terminus of the alignment from Jeffry Hill to Browsholme Heights would have been visible, less than 2 miles (3 km) away. The fact that this correcting alignment appears to have been set out from the north bank of the Hodder, coupled with the fact that it is much easier to sight uphill to a point on the skyline than downhill in the opposite direction, make it likely that this connecting link in the deviation was also set out from north to south. Hence, overall, it appears that this part of the Western Main Road was set out on the ground from both Jeffry Hill and the top of Croasdale, and that the planning exercises met, after some minor correction, near Browsholme Heights.

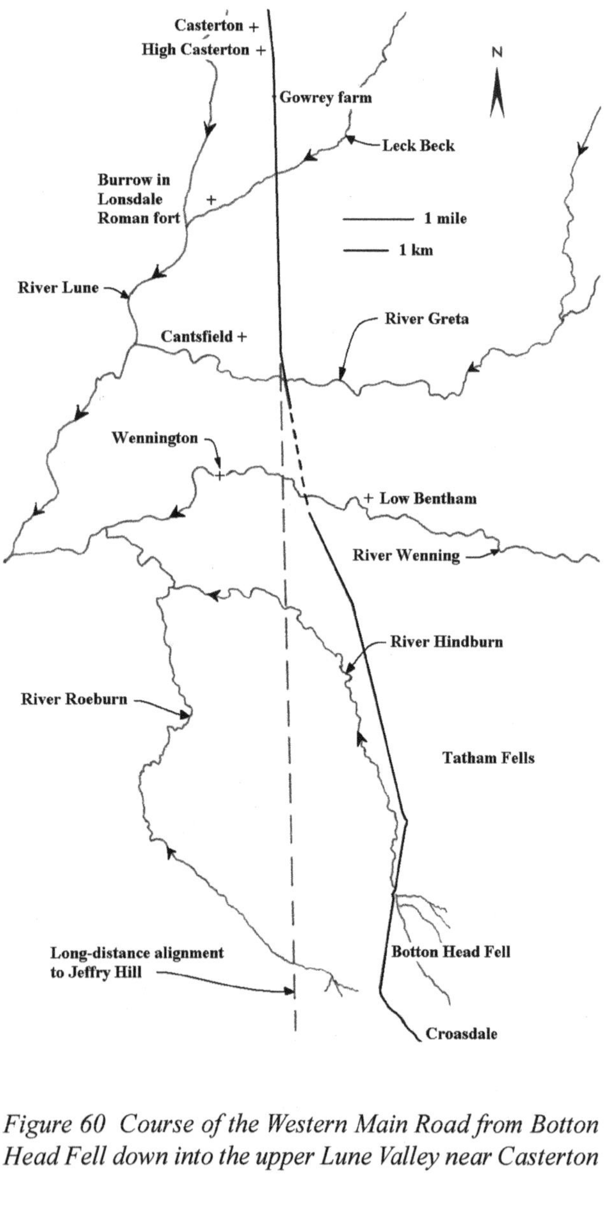

Figure 60 Course of the Western Main Road from Botton Head Fell down into the upper Lune Valley near Casterton

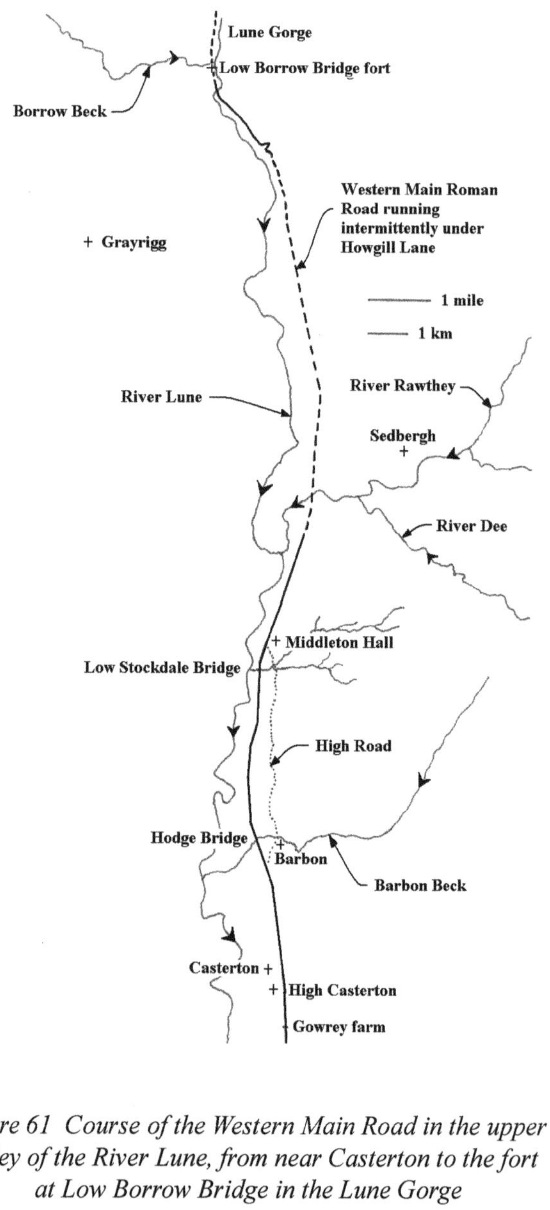

Figure 61 Course of the Western Main Road in the upper valley of the River Lune, from near Casterton to the fort at Low Borrow Bridge in the Lune Gorge

Once the Roman road had climbed over the top of the Bowland Fells on to Botton Head Fell, the intention appears to have been – understandably – for it to lose height as rapidly as possible. This was achieved by a series of straight stretches down Botton Head Fell, across the headwaters of the River Hindburn, and then down past the district of Tatham to cross the River Wenning between Wennington and Lower Bentham. See Figure 60. Probably these stretches were all set out working downhill, from south to north, but this has not been checked on the ground. However, as has already been noted, the way in which the final part of this deviation curves on to the projected planning line from near Casterton suggests strongly that the planning line represented the primary alignment for this stretch of Roman road. Once on the planning line, the course of the road, as built, then followed the alignment exactly to its starting point, near Gowrey farm.

The planning alignment appears to have been set out from a point about 280 yards (256 m) north of Gowrey farm, which lies about ½ mile (0.8 km) south-east of the village of High Casterton. Since the best field of view at this point is south-facing, and the entire long-distance planning for the road in this area appears likely to have been set out southwards from the Lune Gorge, it may be expected that the setting-out of the course of the road on the ground along the Lune Valley could also have been carried out from north to south. Although the direction of planning of the Roman road here has not been verified in detail, there are a number of indications that this might indeed have been the case.

To take the uppermost part of the Lune Valley first, though, it must be observed that between the Lune Gorge and the crossing of the River Rawthey, west of Sedbergh, the course of the Western Main Roman road appears to be followed only loosely by the minor road today which is known as Howgill Lane. See Figure 61. The resulting uncertainty about the exact line of the Roman road in this part of the Lune Valley thus means that no conclusions can be drawn about the direction or directions of its planning there.

South of the River Rawthey, however, the course of the Roman road is shown clearly on Ordnance Survey maps, and it appears to be a series of short straight stretches, most of which are followed by the modern A683 road as far as Hodge Bridge, to the west of the village of Barbon. As already noted, the course of the road here appears to have been set out to run from one ridge of higher ground to another whilst keeping quite close to the River Lune. Nevertheless, along the way, the road is forced to drop down in places to cross streams such as Stockdale Beck and Barbon Beck before they flow into the main river. See Figure 61 again.

What is interesting is that very few settlements lie along the line of this road, as laid out in this way. Nearly all of the farmsteads between Middleton Hall and Barbon lie on or beside a minor road called High Road. As can be seen on Figure 61, this road runs about 500 yards (c.460 m) to the east of the Roman road, and is very narrow and twisty. However, the course of this High Road is surprisingly level, and on the surface it appears that it would have been possible for the Romans to have laid out a better line for their road if they had kept it further away from the river. The fact that the Roman road runs southward straight past Middleton Hall before falling to Stockdale Beck is one of the indicators that this stretch of the Western Main Road may have been set out from north to south.

South of the Barbon Beck the Roman road parts company from the A683 and continues in a series of short straight stretches to rise gently to higher ground above High Casterton before beginning to fall to the point north of Gowrey farm where it appears that the long-distance planning line was set out to run almost due south over the Bowland Fells.

North of the Lune Gorge, the Western Main Road is now known to be heading for the Roman fort at Kirkby Thore rather than the one at Brougham, thanks to the work of Hugh Toller and his colleagues. However, as has already been noted, rather than taking a direct line towards Kirkby Thore through the vicinity of Orton, the Roman road adopts a line slightly to the west of north, seemingly so as to swing on to the upper long-distance survey line from Lancaster that was discovered by Rob Entwistle. See Figure 62. The way in which the Roman road, as built, veers towards this alignment here and then runs onto it can leave little doubt, it is believed, that this line did exist in Roman times, and that it was employed as part of the long-distance planning framework for the Western Main Road. Before the road itself came to be constructed, it is possible that this survey alignment may have been used as a guide for the supply trains supporting the Roman forces campaigning ahead, much in the same way as the planning framework for Dere Street is considered likely to have served initially for guidance on the eastern side of the Pennines. See Section 5 above.

The initial alignment of the Western Main Road northwards from the Lune Gorge, of rather more than 3½ miles (5.6 km), seems to have been set out from Loups Fell, high above the northern exit from the Gorge. The picture is complicated immediately by a short deviation which the road takes, presumably to cross the Birk Beck at a suitable place below Loupsfell Side farm. This deviation returns to the main alignment just north of the motorway service station, approximately where the Roman road is crossed by the southbound carriageway of the M6. As shown on Figure 62, it can then be observed that although Rob Entwistle's survey line rises over a shoulder of Crosby Ravensworth Fell, the main alignment of the road from Loups Fell can be seen to be heading, very sensibly, for a saddle in the hillside to the west of How Nook Pike, before swinging back to run on to the projected survey line at about ¾ mile (1.2 km) before Low Dalebanks farm.

North of Low Dalebanks farm, the survey line would have crossed the River Lyvennet somewhat to the south of

Barnskew and Roans, but here the river is flowing through the centre of a long, wide, flat and low-lying area that looks as if it would be very prone to flooding and thus be quite unsuitable for the course of a road. Not unexpectedly, therefore, the course of the road, as revealed by LiDAR images, shows a deviation that postpones the Roman road's crossing of the Lyvennet until near High Whitber farm. See Figure 63.

What is interesting about this deviation is that instead of being a relatively short one around the low-lying valley floor, it is extended over quite some distance, starting by climbing more steeply than the survey line from Low Dalebanks farm. The route chosen is a practical one, though, and along the way it succeeds in avoiding an awkward little side valley to the west of Meaburn Hall. Thereafter, having crossed the River Lyvennet near High Whitber farm, the road as built does not appear to return to the survey line. This, again, seems sensible since such a detour would look to have offered no advantage over this gently rolling countryside. Instead, the Roman road, as built, appears to have headed more or less directly towards the fort at Kirkby Thore. The final part of its course to Kirkby Thore was probably determined by the selection of a suitable location at which to cross the River Eden as much as by anything else. The extent of this deviation, implying quite extensive local surveying, exhibits an unusual degree of sophistication in the planning of a Roman road at this time.

Beyond Kirkby Thore, it is not clear that the Maiden Way is a continuation of the Western Main Roman Road. See Figure 54. Its construction does not appear to be as substantial as that of the Western Main Road, and at least over it highest point, by Meg's Cairn, no side ditches are to be observed. In addition like its modern counterpart, the A685 road over Hartside Top, which crosses these fells a little further to the north, this Roman road must have regularly become snowed up and impassable every winter. Nevertheless this road appears to make use of the same long-distance survey line as the Western Main Road, and therefore it seems appropriate to include an examination of its planning here.

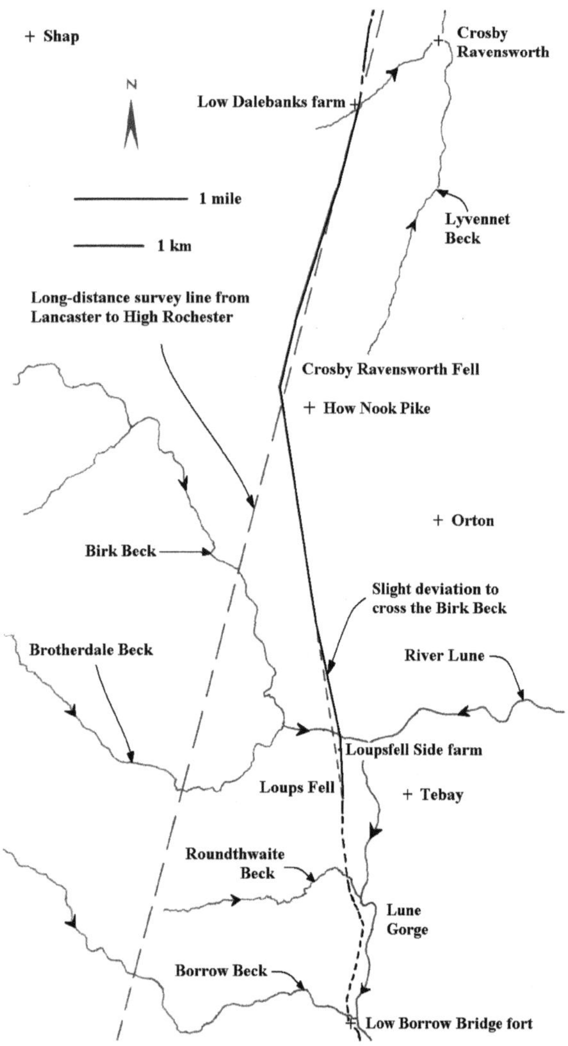

Figure 62 Course of the Western Main Road north of the Lune Gorge, ascending on to Crosby Ravensworth Fell and falling in with the long-distance survey line from Lancaster to High Rochester

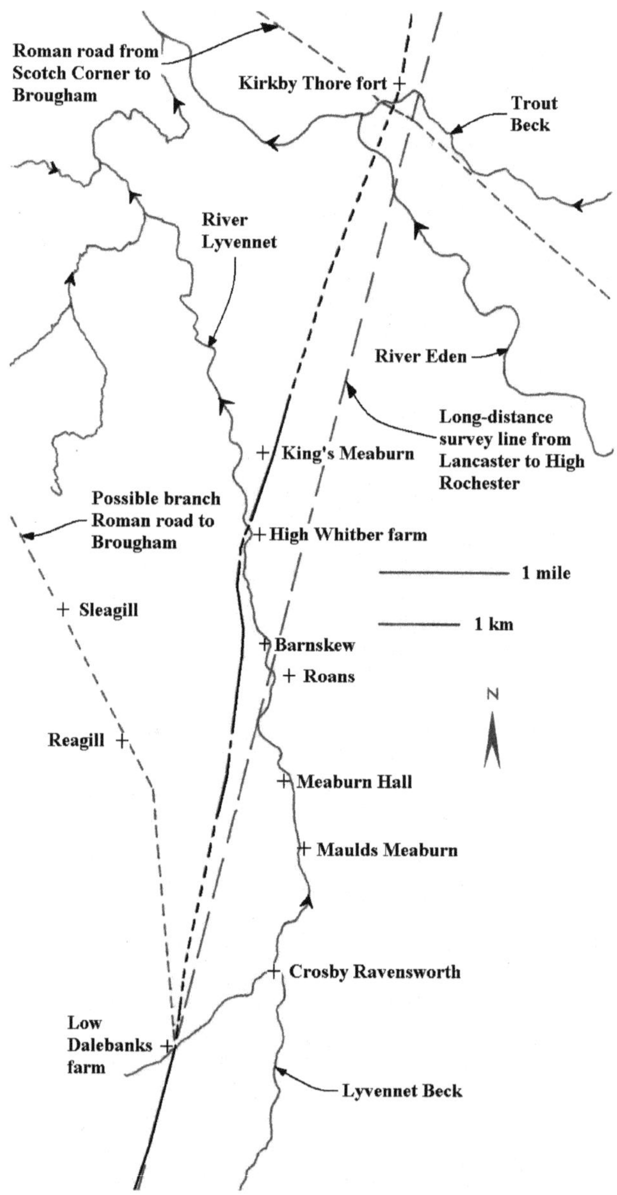

Figure 63 Course of the Western Main Road from near Crosby Ravensworth to the fort at Kirkby Thore

As already observed, the long-distance survey line from Lancaster would have run over the shoulder of Cross Fell, and would have offered a quite impractical route for a road. Instead, after leaving the fort at Kirkby Thore, the Maiden Way appears to have headed directly towards the nearest practical pass through these fells, by Meg's Cairn. Even so, along the way, the Roman road would have had to climb some 1200 feet (c.365 m) in little more than 2 miles (3.2 km), and it is clear that once embarked upon the ascent, the course of the road would have been entirely determined by the topography. In these circumstances, it can be seen on Figure 64 that the road, as built, contrives to maintain a remarkably close allegiance to an alignment upon Meg's Cairn. Such an alignment would have been an entirely practical one to have set out, since the saddle at Meg's Cairn would have been visible from the fort at Kirkby Thore. In fact, Meg's Cairn lies exactly on the skyline when viewed from Kirkby Thore, and this raises the possibility that, contrary to local legend, the origin of the cairn may have been Roman, and that it may have been created by their surveyors to help set out the line that the road should take up to the pass.

Past Meg's Cairn, where, incidentally, the remains of the Maiden Way happen to be very well preserved on the ground, the course of the road continues to be dictated by the topography, the line taken being a gently-curving but predominantly dry one down the ridge between Aglionby Beck and Smittergill Burn. Nonetheless, although quite direct, the course of the Roman road appears to be working back towards the long-distance survey line. See Figure 65. This trend continues until, below Scarberry Hill in the Gilderdale Forest, the road turns on to the line and runs upon it for up to 1½ miles (2.4 km) before bearing away northwards towards the Roman fort at Whitley Castle. On the ground, the course of the road through the Gilderdale Forest appears to be a good one, but there appears to be no particular physical reason why the Maiden Way should have run exactly upon the survey line. Hence this seems unlikely to have been an accidental concurrence.

Beyond Whitley Castle, the general direction of the Maiden Way changes from east of north to slightly west of north, and it is clear that the focus of its planners had shifted to

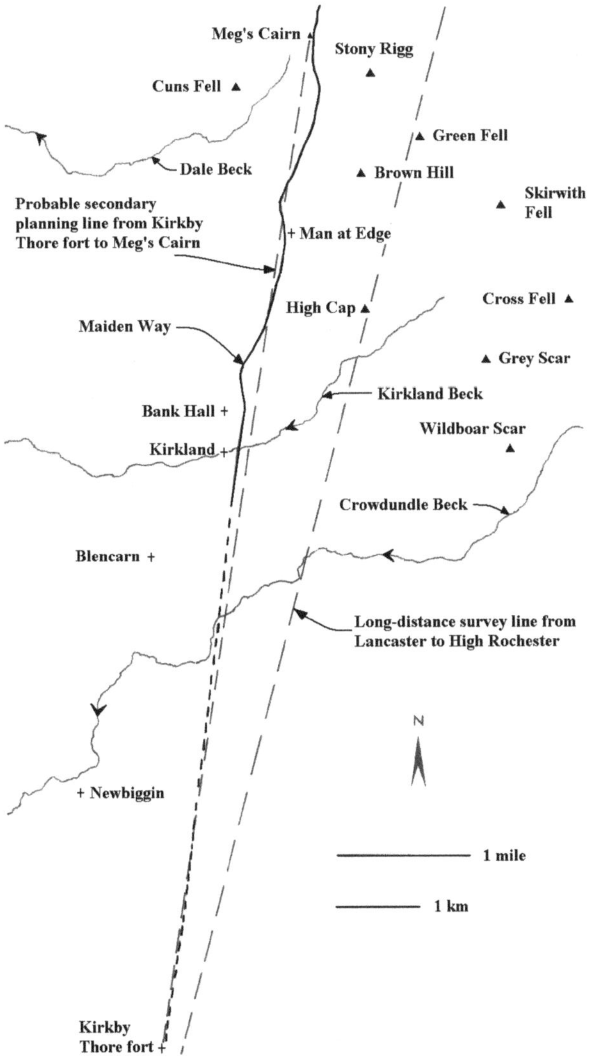

Figure 64 Course of the Maiden Way from Kirkby Thore fort up to Meg's Cairn

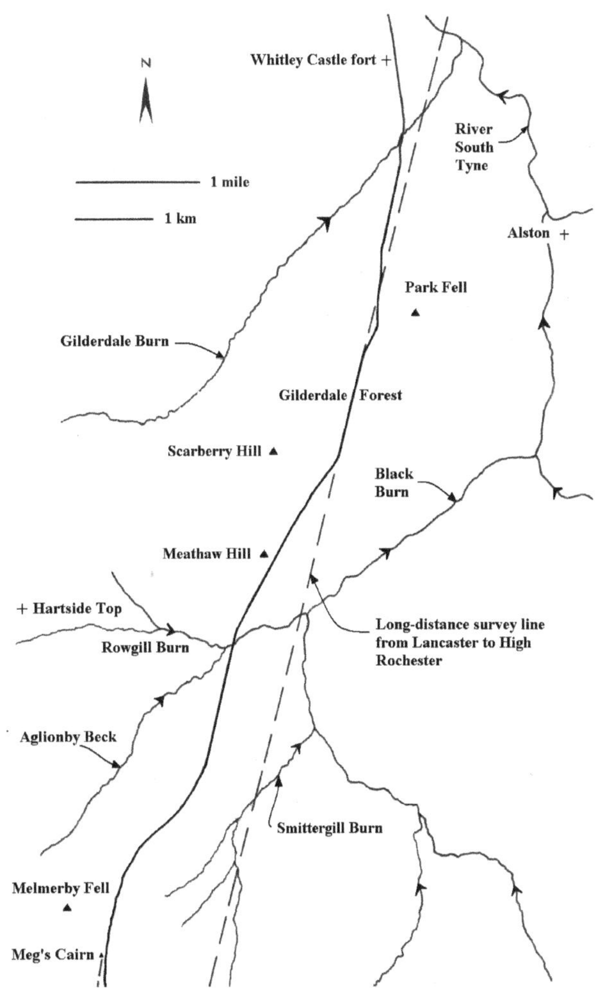

Figure 65 Course of the Maiden Way from Meg's Cairn to the fort at Whitley Castle

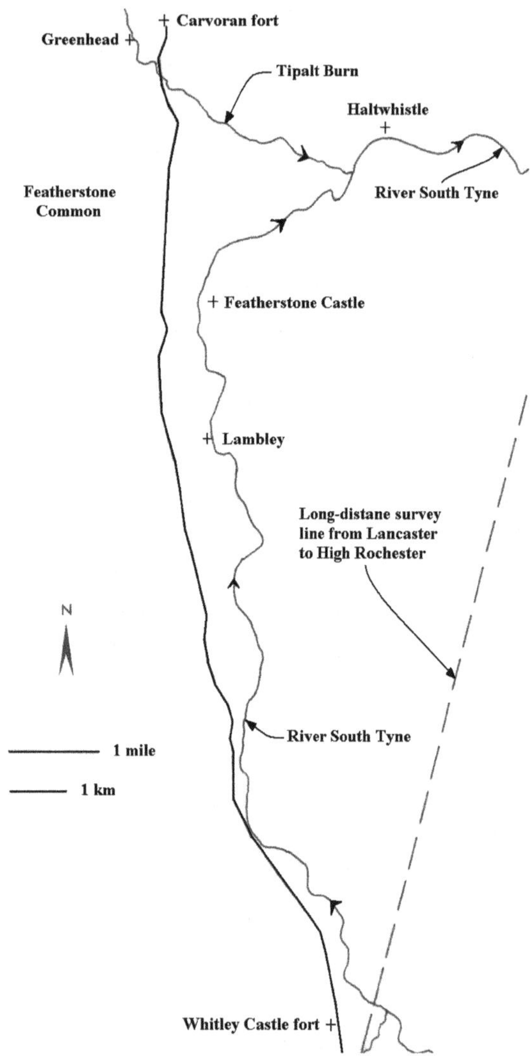

Figure 66 Course of the Maiden Way from Whitley Castle to the fort at Carvoran

following the valley of the River South Tyne as it heads towards its confluence with the Tipalt Burn. See Figure 66. There is no indication that any long-distance alignment was being followed here. Instead, the Roman road proceeds via a number of straight stretches seemingly adapted to the ground, in a manner which is similar to the course of the Western Main Road in the upper Lune Valley. Finally, opposite Featherstone Castle, the Maiden Way adopts a line which quits the South Tyne Valley and heads over Featherstone Common before turning to cross the Tipalt Burn and then run up to the Roman fort at Carvoran. There is just a possibility that the main alignment over Featherstone Common might have been set out southwards from about NY 666 634 by sighting back on to the eastern side of what may have been an earlier and larger fort at Carvoran, but this supposition is not very convincing and it is more likely that this possible alignment was simply a coincidence.

Concluding reflections

It thus appears most likely that the long-distance planning framework for both the Western Main Road and the Maiden Way was set out using existing survey lines that had radiated out from Lancaster, coupled with an exploratory planning line which had been set out over the Bowland Fells from the Lune Valley. Figure 67 shows this perceived overall long-distance planning framework, and Figure 68 shows the Western Main Road and the Maiden Way as built in relation to that framework. As with the long-distance framework for Dere Street, it is possible that the two long-distance survey lines here may, in part at least, have been used initially to help guide supply trains and communications links in support of the Roman forces campaigning further to the north and east. However, unlike the case with Dere Street, no evidence has yet come to light along the Western Main Road to underpin this supposition.

Nonetheless, like Dere Street, it would appear that the likeliest period for the setting-out of these lines would have been during or in the immediate aftermath of the Roman conquest of the area, when it can be presumed that the dispositions of the forts and roads were being determined by the Roman military authorities for the long-time occupation of the region. For reference, the earliest periods of occupation which are currently deduced for the known forts between Manchester and High Rochester are described below.

Manchester	Occupation began under Agricola, or possibly a few years earlier[72]
Ribchester	Before Agricola and probably as early as the governorship of Cerialis[73]
Burrow in Lonsdale	The northern rampart and ditch appear to have been of Flavian date[74]
Low Borrow Bridge	There is pre-Hadrianic pottery from the site, and a large ditch running inside the line of the western defences was perhaps part of an earlier fort[75]
Kirkby Thore	Samian ware suggests an Agricolan date for the establishment of the fort[76]
Whitley Castle	Pottery recovered dated from the mid 2nd century, but excavation of the fort has been very limited.[77] On the basis that the fort would have been located to oversee Roman mining operations in the area, and noting the rapidity with which the Roman army took control of significant mineral resources elsewhere in Britain, Went and Ainsworth suggest that the fort may have been established at around the turn of the first century, or at least no later than the reign of Hadrian.[78] Admittedly, however, this is speculation
Carvoran	Ditches revealed early second century (Trajanic) pottery, but little excavation has so far taken place at this site[79]
High Rochester	The site was occupied in the Flavian period[80]

With regard to the question of when the detailed planning and construction of the Western Main Road and the Maiden Way took place, it may be helpful to examine the situation on the eastern side of the Pennines. There, study of the course of Dere Street indicated that the detailed planning and construction of that Roman road is likely to have taken place some years after the creation of its long-distance framework and to have been conducted in piecemeal fashion. Thus, from the available evidence, different parts of Dere Street (away from the immediate vicinity of forts) seem to have been constructed in the main between AD 85 and AD 105.[81] Although such evidence is lacking for the Western Main Road and the Maiden Way, a similar timescale can be envisaged for them as well, albeit that, from the foregoing dating evidence it is conceivable that the Maiden Way may have been constructed rather later than the Western Main Road.

The point is that, just as with the creation of the motorway network in Britain from the 1950s, it can be imagined that the date and location of any construction would have depended upon priorities at the time, and so piecemeal construction should not be unexpected in Roman times, especially over a total length of Roman road (including the Maiden Way) of some 110 miles (177 km).

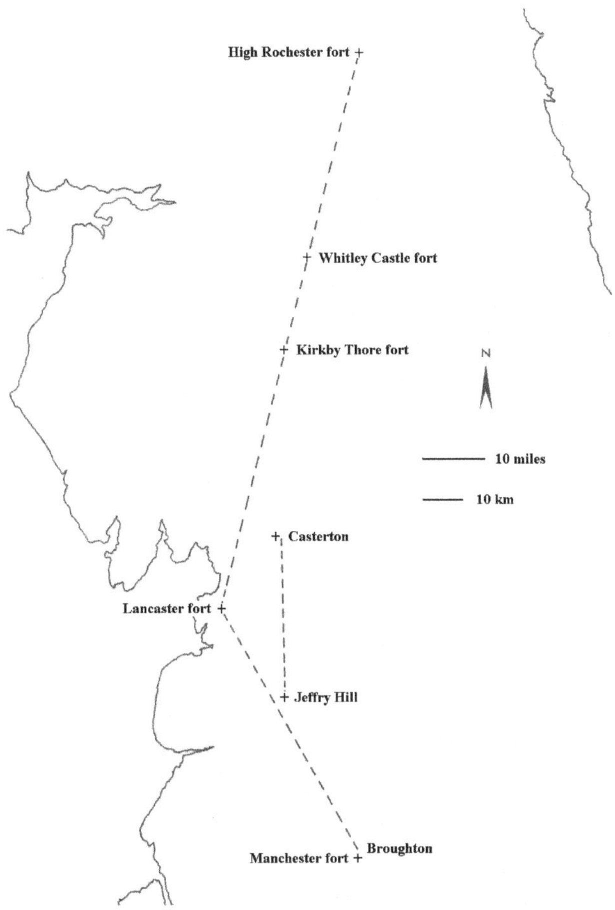

Figure 67 Perceived overall long-distance planning framework for the Western Main Roman Road and the Maiden Way

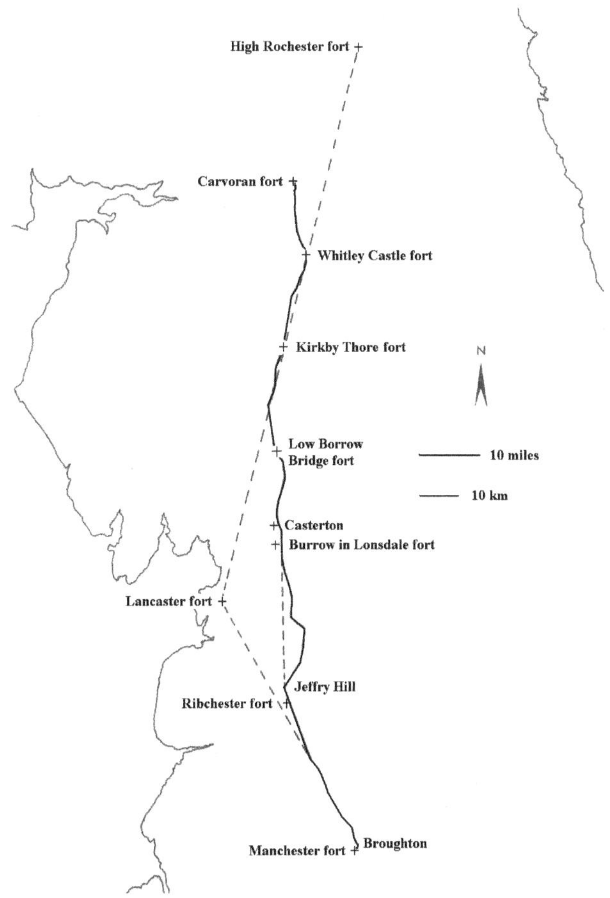

Figure 68 Courses of the Western Main Roman Road and the Maiden Way, as built, in relation to the overall long-distance planning framework

[72] Bidwell and Hodgson, 2009, 97
[73] *Ibid*, 99
[74] *Ibid*, 101
[75] *Ibid*, 102
[76] *Ibid*, 64
[77] *Ibid*, 128-30
[78] Went and Ainsworth, 2013, 130
[79] Breeze, 2006, 280
[80] Bidwell, 2009, 4; see also Hodgson, 2009, 168; and Hancke, Charlton, and Biggins, 2004, 42.
[81] Poulter, 2010, 38-46

Section 8: Example of the surveying and planning of the network of cross-country roads from Kirkham to Aldborough via Ribchester, Elslack, and Ilkley

In northern England, in addition to the two trunk roads – the Western Main Road and Dere Street – which ran south to north on either side of the Pennine Hills, the Romans also laid out a number of roads across the Pennines, thus forming something akin to a ladder-like network. See Figure 1. Naturally, gaps in the Pennine chain were used for these cross-country roads whenever possible. Thus the Stanegate road runs across the low-lying land between Corbridge and Carlisle; the road from Scotch Corner past Kirkby Thore to Brougham, near Penrith, uses the Stainmore Pass; and the road or network of roads from Kirkham to Aldborough runs through the Craven Gap.

Ivan Margary and earlier commentators seem to have assumed that the road which passed through the Craven Gap ran from Ribchester to Ilkley and then on via Tadcaster to York. Thus Margary allocates the Ribchester to Ilkley road his number 72a,[82] but gives different numbers to the roads between Ribchester and Kirkham (703)[83] and Ilkley and Aldborough (720b).[84] However, in terms of planning at least, it looks as if the sequence of roads spanning the 64½ miles (103.3 km) from Kirkham to Aldborough could have been based upon a single alignment.

Long-distance planning

It was, again, Rob Entwistle who pointed out that the Roman forts at Kirkham and Elslack, and the civitas capital at Aldborough, all lay on a single alignment, as shown in Figure 19, and he noted that this alignment was also followed by the Roman road from Downham Park, near Clitheroe, to Skipton, as shown on Figure 69. Statistically, the fact that three Roman sites lie on the same line may not be significant, but the way in which the Roman road turns abruptly on to the alignment in Downham Park and then follows the line, albeit somewhat loosely in places, for some 14½ miles (c.23 km) makes it very likely that this alignment did exist in Roman times and that it was used to help lay out the course of the road. The rather unusual way in which the central stretch of this road, between Downham Park and Skipton, slips off the alignment in places will be examined in detail later, but for the most part the detours appear to be explicable by the topography.

There appear to be two main but alternative explanations of how this long-distance alignment could have been set out. Both seem to be equally possible, but neither is entirely satisfactory.

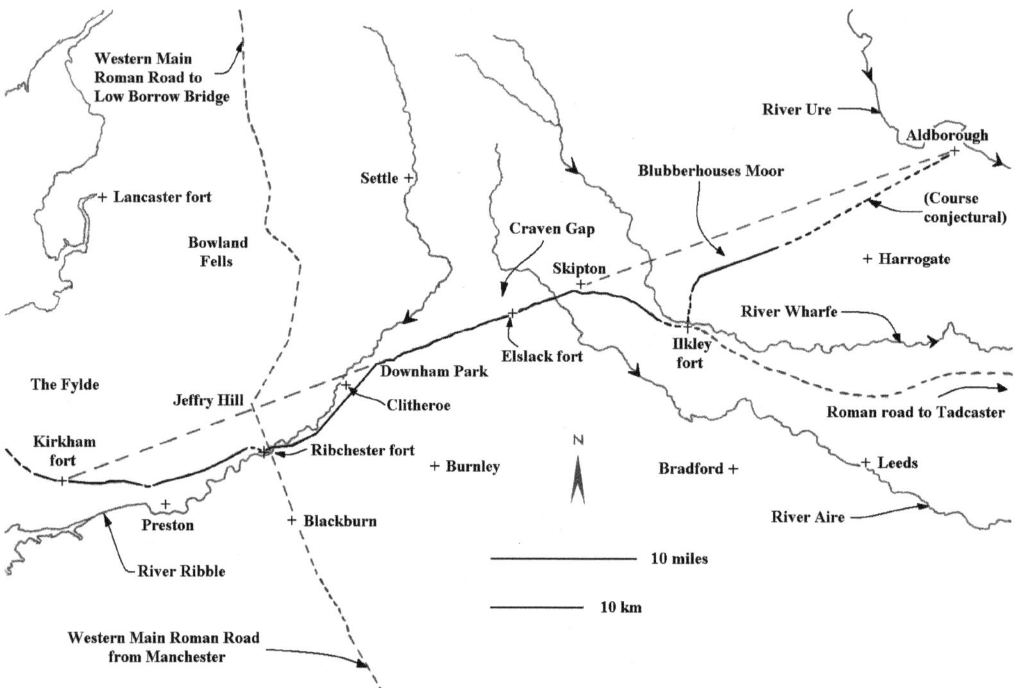

Figure 69 Course of the cross-country Roman road from the fort at Kirkham through the forts at Ribchester, Elslack, and Ilkley to the civitas capital at Aldborough

[82] Margary, 1973, 371-4
[83] *Ibid*, 374-5
[84] *Ibid*, 406-7

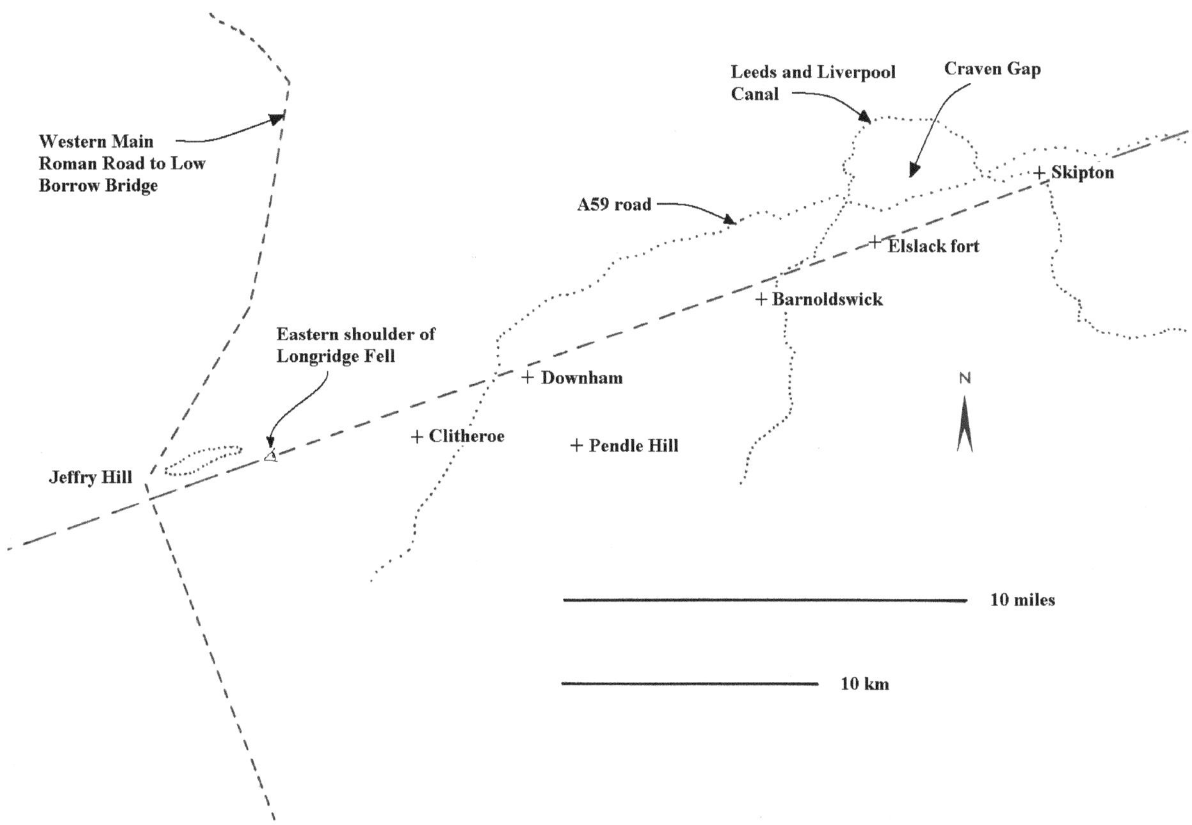

Figure 70 Possible sighting through the Craven Gap from the eastern shoulder of Longridge Fell

The first possibility is that the alignment had initially been set out by the road planners in order to plot a route through the Craven Gap, and that the line had then been simply extended to the west and the east to help fix the locations of Kirkham and Aldborough. In the case of Kirkham, this would have been when a suitable site for a fort had been encountered along the line, once it had been extended westwards into the area of the Fylde, east of Blackpool. In the case of Aldborough, at the eastern end, it would have been when the line reached the River Ure.

The Craven Gap is used today by the A59 road between Clitheroe and Skipton, and also by the Leeds and Liverpool Canal. However, in passing from west to east, the Gap curves a little to the north, and if the Romans, in typical style, had sought to set out a straight line through the Craven Gap and across the intervening territory, they would have needed to get as good a view as possible into and through the Gap itself. A suitable position from which to have done this would have been from a point at about SD 684 410 on the south-eastern shoulder of Longridge Fell. See Figure 70. Unfortunately, it is not easily possible to test this supposition entirely at present because this shoulder of the Fell is heavily wooded. However, from the foot of the tree line, looking eastwards, the notch marking the start of the Craven Gap can be made out, indicating that, without the hindrance of the modern-day tree cover, setting out an alignment through the Gap from the higher sighting point would appear to have been an entirely practical affair.

It is considered highly unlikely that an alignment through the Craven Gap could have been set out by sighting from either end point. The Craven Gap is completely invisible from both Kirkham and Aldborough, and there is nothing which might indicate its possible presence along either of their skylines. The summit of Longridge Fell is clearly visible from Kirkham, but a line from there to the summit would have run well to the north of the Craven Gap.

The alternative possibility is that the line through the Craven Gap may have been derived from preparations, at least, for a scheme of centuriation that had been intended for the Vale of York. It might have been found that a line parallel to one of the axes of the centuriation happened to pass through the Craven Gap on a suitable alignment, so that this line was then used to lay out the course of the road and also, by extension, to fix the locations of Aldborough, Elslack, and Kirkham. Certainly, whilst Kirkham and Elslack look to be suitable sites for forts, the settlement at Aldborough has the appearance of having been chosen by bureaucrats, lying on a sloping hillside running down to the flood plain of the River Ure.[85]

What elevates this possibility above the level of pure speculation is the course of the Roman road from Ilkley to Aldborough. On Blubberhouses Moor, having run

[85] If, however, the civilian settlement at Aldborough had been preceded by a fort, as is strongly suspected, then this fort might have been sited on lower-lying land closer to the river. The author is grateful to Mike Haken for this suggestion.

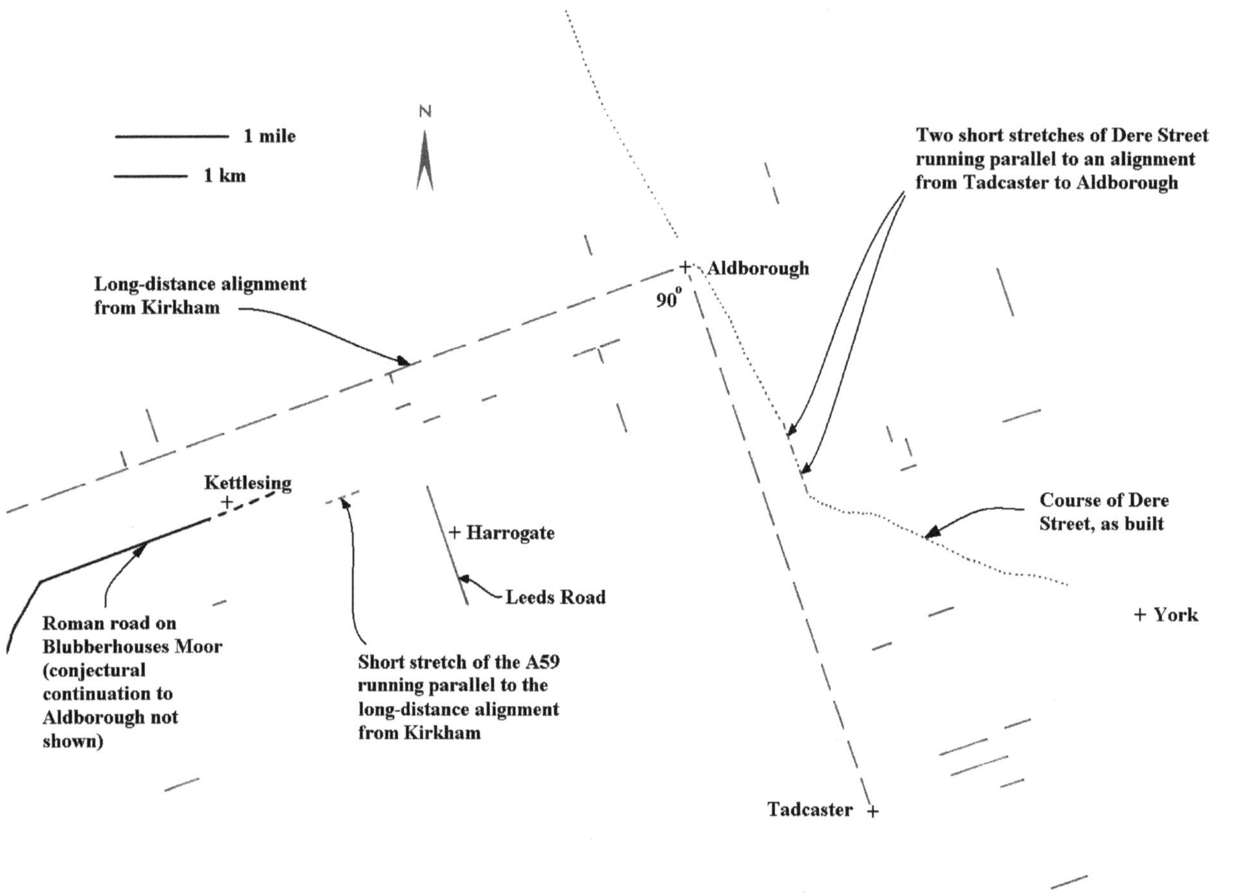

Figure 71 Indications of a possible scheme of centuriation (limitatio) in the Vale of York. On balance the evidence is considered insufficient to be able to conclude that – even if such a scheme had been started – it had ever been completed

northwards from the fort at Ilkley and having then reached the shoulder of Round Hill, the road turns abruptly and runs along an alignment which is exactly parallel to the alignment from Kirkham to Aldborough. See Figure 71. No practical reason for such a parallel alignment can be detected on the ground. Indeed, this would have been a laborious as well as a seemingly pointless feat to have achieved if the alignments had not existed already. As it happens, the two alignments are close to 90 *actus* apart,[86] a round number which might appear to give further encouragement that this feature is not a coincidence.

There are other indicators that such a scheme of centuriation might have been afoot in the area. As noted in Section 5, Tadcaster appears to have been the starting-point (or possibly a way-point) for the long-distance planning framework for Dere Street. It is conceivable that, subsequently, Tadcaster may have been used as the base for a scheme of centuriation in the area. This is because a direct line from Tadcaster to Aldborough lies at very close to a right angle to the Kirkham to Aldborough alignment.[87] See Figure 71 again. Moreover, two stretches of Dere Street, as built, from SE 456 571 to SE 453 578, and from SE 449 588 to SE 446 597, appear to run parallel to the line from Tadcaster to Aldborough, and at about 49 *actus* from it – close to a round figure of 50 *actus*, but perhaps not enough to be persuasive.

In addition to the foregoing indicators, a short stretch of the modern A59 road from Skipton to Harrogate appears to run parallel to the Kirkham to Aldborough alignment near Saltergate Hill. This lies further east than the rather longer stretch of the A59 which is known to follow the alignment of the Roman road towards Kettlesing after it had descended from Blubberhouses Moor and crossed the Fewston Reservoir. Further, the Leeds Road in Harrogate, although not known as a Roman road, runs parallel to the Tadcaster to Aldborough alignment at a distance of some 380 *actus*. However, it is possible that neither of these stretches of road had existed before the 18th century,[88] and there are few other indications of a centuriated landscape in the area.[89] Finally, to round matters up, after the results were discussed with John Peterson it was concluded that

[86] An *actus* is a Roman unit of linear land measurement, equal to 120 Roman feet. Each grid in a scheme of Roman centuriation might, typically, be 20 actus square. It terms of modern measurements, an *actus* equates to around 38.8 yards (35.5 m). Thus, in a scheme of centuriation, a typical grid would be about 776.5 by 776.5 yards (710 m by 710 m).

[87] The angle appears to be 90.9°.

[88] Mike Haken, *pers. comm.*

[89] In 1980, Herman Ramm drew attention to possible indications of a centuriated landscape lying to the west of Tadcaster; see Ramm, 1980, 33-5. However, the evidence presented failed to impress John Peterson (*pers. comm.*) as being sufficiently orthogonal.

more evidence would have been expected if such a scheme of centuriation had been completed in the area.[90]

On balance, therefore, of the two possible models for the planning of the Kirkham to Aldborough alignment, the author is more inclined to believe the former: i.e. that the line was first set out in order to plot a course for the Roman road through the Craven Gap. Quite apart from the uncertainties about a possible scheme of centuriation in the area, to the author, having examined the landscapes on the ground, the former just seems to be the more practical model. It is not that the Craven Gap would present any great difficulty for road construction, but that the nature of the land in the Gap consists of a warren of small hills and valleys in which it is easy to lose one's bearings. In this respect the Craven Gap is not unlike the low-lying land across which the Rivers Greta, Wenning and Hindburn flow to join the southern arm of the River Lune to the west of Cantsfield and Bentham (see Figure 60). Hence, in both of these cases, setting out a line to plot a direct course for a road through landscapes like this would appear to make sense.

This may also help to explain why, in the case of this network of roads, the line deviates quite substantially from a direct alignment between the two end points, Kirkham and Aldborough. If the alignment had primarily been set out to plot a course through the Craven Gap, then the value of the alignment elsewhere could well have been secondary, at least to the planners of the roads. However, it has to be admitted that this leaves questions unanswered about what was happening on Blubberhouses Moor, which seems highly unlikely to have been a coincidence. In addition, of course, like all interpretations based upon incomplete evidence, both models could be wrong. At any rate, the long-range planning for this network of cross-country Roman roads appears to have been based upon a single alignment, and attention can now be turned to examining the planning of these roads upon the ground in more detail.

Examination of some details of the setting-out of the roads

To begin with the western portion of this network, this may be taken as commencing where the Roman road, as built, turns off the long-distance alignment in Downham Park. If the road had continued westward along the alignment without turning it would have reached the River Ribble above Brungerley Bridge, towards Waddington, at something of a cliff. See Figure 72. However, this obstacle could easily have been circumvented by a short deviation either upstream or downstream, after which it appears that a suitable course for the road could have been laid out without difficulty down the northern side of the Ribble to the Roman fort at Ribchester.

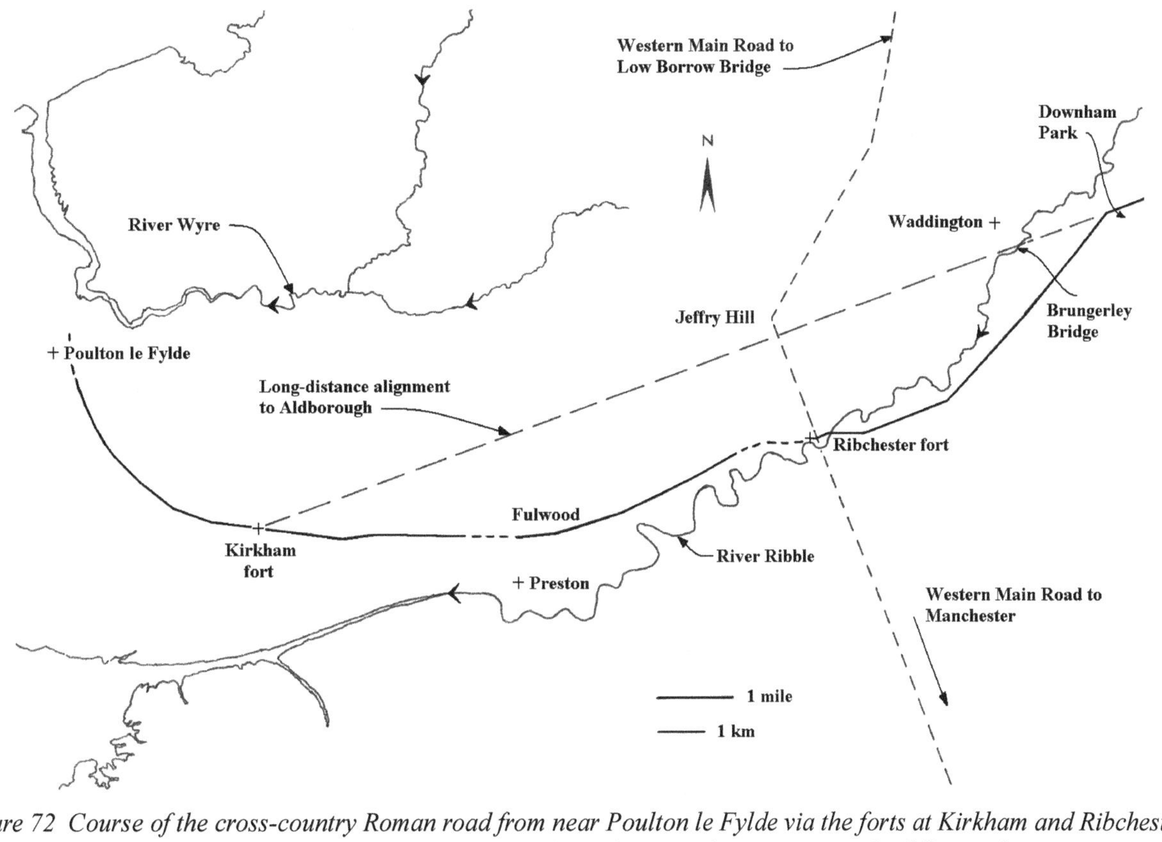

Figure 72 Course of the cross-country Roman road from near Poulton le Fylde via the forts at Kirkham and Ribchester to the turn, in Downham Park, onto the long-distance alignment towards Aldborough

[90] John Peterson *pers. comm.* In addition to this verdict, Brian Campbell (*pers. comm.*) commented that the distance of 90 *actus* between the alignments on Blubberhouses Moor does not fit into any pattern of measurement in the surveying literature.

Instead, by turning off the long-distance alignment in Downham Park, the road, as built, adopted a course that kept it to the south of the River Ribble until, it appears, it crossed the river by the De Tabley Arms, about ½ mile (0.8 km) to the east of the Roman fort. Delaying the crossing of the River Ribble until reaching the vicinity of the fort would appear to make sense militarily, but it implies that the location of the Roman fort had been decided before the course of this cross-country road had been set out.

The location of the Roman fort at Ribchester appears most likely to have been determined when the long-distance framework for the Western Main Road was being worked out. Interestingly, if both the Western Main Road and this cross-country road had crossed over the Ribble by bridges at the locations to which their lines point today, this would have meant that the Roman fort would have had two crossing-points of the river to protect, little more than ¼ mile (0.4 km) apart. See Figure 72 again. Militarily, this duplication would also appear to have made sense, however, since, like most major rivers in the north of England, the Ribble can be prone to widespread and destructive flooding at times. It can therefore be imagined that one or other of the bridges over it near Ribchester would have been rendered unserviceable and would need to have been repaired or replaced on more than one occasion during the Roman occupation.

West of Ribchester the course of the cross-country Roman road exhibits no inclination to return to the long-distance alignment, its planners seemingly preferring its line to remain closer to the Ribble along its northern flanks. As shown on Figure 72, this – albeit somewhat distant – allegiance to the river is retained through Fulwood, in the northern suburbs of Preston. After this point, however, with the river continuing in a south-westerly direction towards the sea, a series of gradual turns more northwards was adopted to bring the line of the Roman road back to the long-distance alignment at the fort at the Kirkham. The line of the road is continued beyond Kirkham, curving ever more northwards before its course is now lost towards the east of Poulton le Fylde. Margary suggests that the destination of this road might have been a harbour on the River Wyre,[91] and this would seem to be a reasonable presumption.

To turn now to the central part of this cross-country Roman route, this is where it appears to follow the long-distance alignment for some 14¼ miles (c.23 km) from Downham Park through the Craven Gap to Skipton. In so doing, the most notable feature of the road is the looseness with which the alignment is followed in places. This is both striking and curious because elsewhere along its course – such as south-westwards from Downham Park past Clitheroe – the planners of this road appear to have had no difficulty in following straight alignments exactly. It may have been that, between Longridge Fell and Skipton, the long-distance alignment had been only sparsely marked out – for instance, only from high point to high point, such as on Longridge Fell and by Great Todber farm, as indicated in Figure 73.

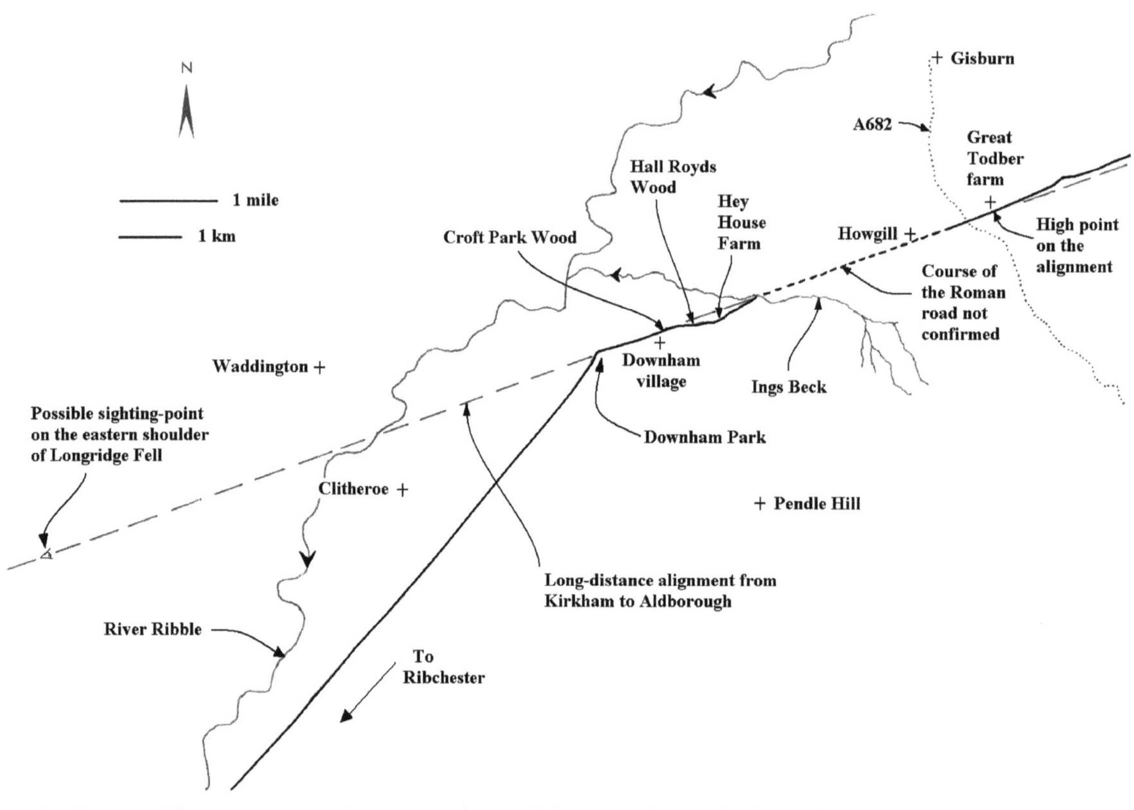

Figure 73 Course of the cross-country Roman road past Clitheroe and on to the long-distance alignment in Downham Park

[91] Margary, 1973, 375

No such straightforward explanation would suffice, however, for the first detour that the Roman road makes after quitting Downham Park. After having followed the long-distance alignment for rather more than ¾ mile (1.2 km) through Croft Park Wood to the north of Downham village, the Roman road abandons the alignment to follow the summit of a local ridge past Hall Royds Wood and then to run down via Hey House Farm to rejoin the long-distance alignment in the vicinity of the crossing of the Ings Beck. See Figure 73 again. This is one of the most pronounced detours which the Roman road makes from the long-distance alignment between Downham and Skipton, and there appears to be no physical reason for it that can be seen nowadays. The course of the long-distance alignment here would have been an entirely practical one for the road to have taken, but an inspection of that line on the ground to the north of Hall Royds Wood revealed no sign of any road construction along its length. Instead, the course of the Roman road can clearly be seen on the ground to run where Margary[92] and, earlier, Percival Ross[93] described it to be, and where the Ordnance Survey has also shown it to be on their 1:25,000 (2½ inches to the mile) Outdoor Leisure Map.[94]

From near where the road turns off the alignment north of Downham village, at SD 788 446, both of the nearest high points – on Longridge Fell and by Great Todber farm – would have been clearly visible. Therefore, for the planners of the Roman road, there should have been no doubt about the course of the long-distance alignment here. Perhaps a clue to the likeliest explanation for this detour lies in the way in which the road, as built, keeps to the highest point of the local ridge. The summit of this ridge lies alongside Hall Royds Wood, and this could have made an excellent location for an observation and signal tower, with fine views to the north, west, and south. It may therefore be that the planners of the road had set out this detour in order to service a tower located at this point. If so, it would appear to have been unusual for a Roman road to have made a detour such as this. More normal Roman practice would appear to have been for the road to keep to its line, with branch roads, if necessary, being constructed to service such signalling stations. Along the Gask Ridge in Scotland, for instance, where several observation and signal towers have been found to lie on either side of the Roman road, the road there follows its alignment without deviation.[95]

Eastwards beyond Ings Beck, the course of the Roman road becomes less certain and then lost altogether as it proceeds towards Howgill. Where the Roman road finally emerges into view again, to the east of Howgill, it can be seen to lie exactly on the long-distance alignment once more, running along the north side of Howgill Lane as the latter rises up to meet the A682 road between Barrowford and Gisburn.

[92] *Ibid*, 372
[93] Ross, 1921, 38-9
[94] Ordnance Survey, 1996
[95] However, the hypothetical possibility that the road along the Gask Ridge had predated the observation and signal towers cannot be entirely ruled out, on present evidence – see Woolliscoft and Hoffmann, 2006, 235.

About ¼ mile (0.4 km) east of the A682 road, the Roman road surmounts one of the high points along the long-distance alignment, near Great Todber farm. Progressing eastwards from this point the presumed sighting point on the eastern shoulder of Longridge Fell would no longer have been visible, and almost immediately the line of the road, as built, begins to diverge from the long-distance alignment. See Figure 74. It may have been that the next markers of the long-distance alignment to the east of this point may not have been readily visible – at least on the occasion when the road happened to be being laid out. These markers could have been either on a high point just east of Thornton in Craven, some 5 miles (8 km) distant, or else on an elevated point by High Skibeden farm to the east of Skipton, some 12 miles (c.19 km) away. See Figure 75. Both of these high points, however, would have had the large mass of Blubberhouses Moor standing higher behind them when viewed from Great Todber farm, and this could have made any markers on them difficult to discern, at least in daylight.

In such circumstances, if the planners of the Roman road had been working from west to east, they may have tried to use any marker by Great Todber farm as a back sight, and so fallen foul of the risk of diverging from the alignment, as explained in Section 6. Alternatively, if the planners of the Roman road had been working from east to west and had strayed off the line before reaching Great Todber farm, then by sighting on any marker by the farm they would have tended to converge on to the alignment as they approached the hill top. As indicated on Figure 74, such convergences towards the alignment may have occurred at SD 861 473 and again at SD 853 470. With so many unknowns about what may or may not have been present in Roman times, it is not possible to proffer this explanation with a high degree of confidence, but it does seem to fit the evidence which can be inspected on the ground today.

Not all the detours of this road lack physical explanation on the ground, however. To the north-east of Coverdale farm, near SD 849 469, the Roman road makes a short detour even further north. On the ground it can be seen that a spring rises on the line that the Roman road would have taken without the short detour. See Figure 74 again. To have attempted to build the road over the spring would have caused unending problems, and hence it appears that the planners of the Roman road had diverted its course a little to the north to a point just downstream of where the flow from the spring joins the stream running down from Coverdale farm. To judge by the ramp for the road on the eastern side of this crossing point, the stream would have been bridged here in Roman times.

Eastwards beyond this point, the Roman road continues to proceed in a series of short straights lying up to 110 yards (c.100 m) to the north of the long-distance alignment. These deviations do not appear to be post-Roman, since the modern successor to the Roman road – Brogden Lane – lies beside rather than on top of the Roman road, and the remains of the Roman road can often be observed from it.

Interestingly, Brogden Lane tends to run closer to the long-distance alignment than does the Roman road, and with one exception there appear to have been no insurmountable physical obstacles which would have prevented the Roman road, as built, from following the alignment exactly. The one exception might have been the steep-sided valley of Ray Gill, which could have presented problems. Overall, it may be that the planners of the Roman road, having temporarily lost track of the long-distance alignment here, had simply resorted to setting out the course of their road to ease the labour and/or increase the speed of its construction. If so, this would have been more akin to modern concepts of road planning and construction.

Whatever the explanation, the somewhat winding course of the Roman road, as built, seems to have continued in this manner past Barnoldswick and Thornton in Craven, and since the Roman fort at Elslack appears to lie exactly upon the long-distance alignment, another detour would have been required to pass by that station. In fact, excavation has indicated that the Roman road seems to have passed close by the fort on its south-eastern side.[96] See Figure 75.

The Roman road is then believed to have continued in a similar vein eastwards from Elslack to cross the River Aire and then run into Skipton along Broughton Road and then Newmarket Street, which now together form part of the A6069 road. From Newmarket Street the Roman road finally and firmly breaks away south-eastwards from the long-distance alignment to run up Shortbank Road and on to the shoulder of Skipton Moor, seemingly so as to begin to cross from Airedale into Wharfedale to service the Roman fort at Ilkley.

The final part of the climb from Shortbank Road on to the shoulder of Skipton Moor is extraordinarily steep, after which the character of the road changes completely. Rather than, in typically Roman style, following a straight line almost regardless of the contours, the road turns sharply to the north of east again and then pursues a curving course on a gently rising gradient along the north face of Skipton Moor. See Figure 75 again. At first sight this looks unlikely to be a Roman road. Moreover, on the surface of the road there are abundant signs of relatively recent road construction, probably no more than 200 to 300 years old. What makes it appear likely that this road had a Roman origin are (a) the very substantial formation of the road underlying the more recent features, and (b) the almost ridiculously steep climb up to it. More than one hundred years ago, Percival Ross had reached a similar conclusion based upon seemingly similar reasoning.[97]

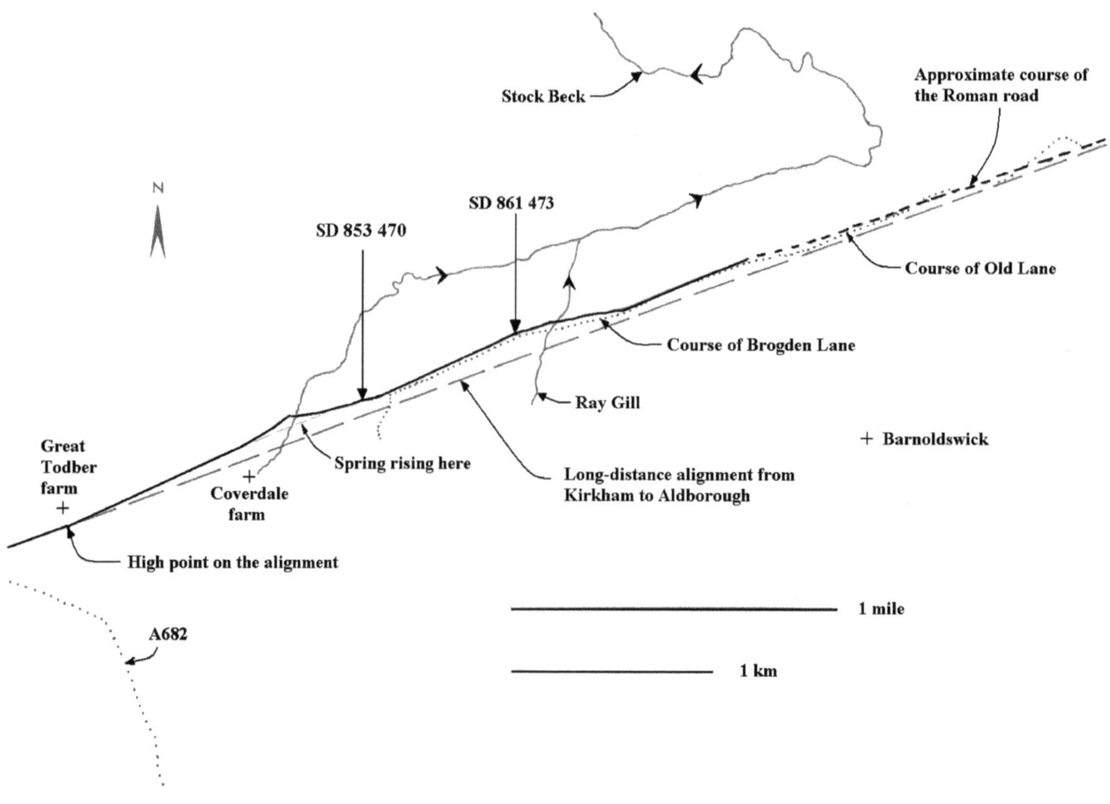

Figure 74 A close-up view of the course of the cross-country Roman road past Great Todber farm and then alongside Brogden Lane to the north of Barnoldswick, and then its continuation either under or alongside Old Lane towards Thornton in Craven

[96] Bidwell and Hodgson, 2009, 104-5

[97] See Ross, 1921, 33-64

Beyond Skipton Moor, this road, having curved to the south-east once more, and by now showing signs of Roman-like alignments, appears to have run into Wharfedale past the village of Addingham and on from there to the Roman fort in Ilkley. See Figure 76. After leaving the fort at Ilkley the Roman road to Aldborough then worked its way northwards, seemingly without following any long-distance alignment. Rather, it gives the appearance of seeking to return to the Kirkham to Aldborough alignment as directly as possible across the intervening moorside, the conjunction with which it would probably have accomplished somewhat to the west of what is now Thruscross Reservoir.

Instead, as already mentioned, some 2½ miles (4 km) short of this potential meeting-point, the Roman road turns abruptly east-northeastwards on the shoulder of Round Hill and then follows an alignment which is exactly parallel to that of the Kirkham to Aldborough alignment for a distance of at least 5¾ miles (9.25 km). If this new alignment had been created as part of an exercise in centuriation, it can be expected that it would have been closely marked out on the ground, at intervals of about 776.5 yards (710m). Certainly, in comparison with the road's somewhat wandering course through the Craven Gap, the planners of this cross-country Roman road now seem to have been able to follow the new alignment much more closely. Even so, the road does gently veer up to 25 yards (c.23 m) to the north of the new alignment when descending across Blubberhouses Moor, before picking up on the alignment more precisely where the valley of the River Washburn would have been crossed. This valley is now occupied by the Fewston Reservoir.

The A59 road from Skipton to Harrogate then picks up on this same alignment and follows it for just over 2 miles (3.2 km) before turning off south-eastwards towards Harrogate. The Roman road is now known to have continued further along the same alignment for some distance, however, and then, beyond the village of Kettlesing, to have turned towards the village of Hampsthwaite. Work is continuing on establishing the line of the road beyond this point, but the presumption is that it was heading for Aldborough.[98] If the behaviour of the road at its western end may be taken as an example, it is possible that this cross-country Roman road, as built, did not finally regain the original long-distance alignment until it reached its terminus at Aldborough. This, however, must remain speculation for the present.

Concluding reflections

As indicated previously, it appears unlikely that the course of this cross-country Roman road to the west of Downham Park would have been laid out before the site of the Roman fort at Ribchester had been determined. However, this leaves the question of when the original long-distance alignment between Kirkham and Aldborough had been set out.

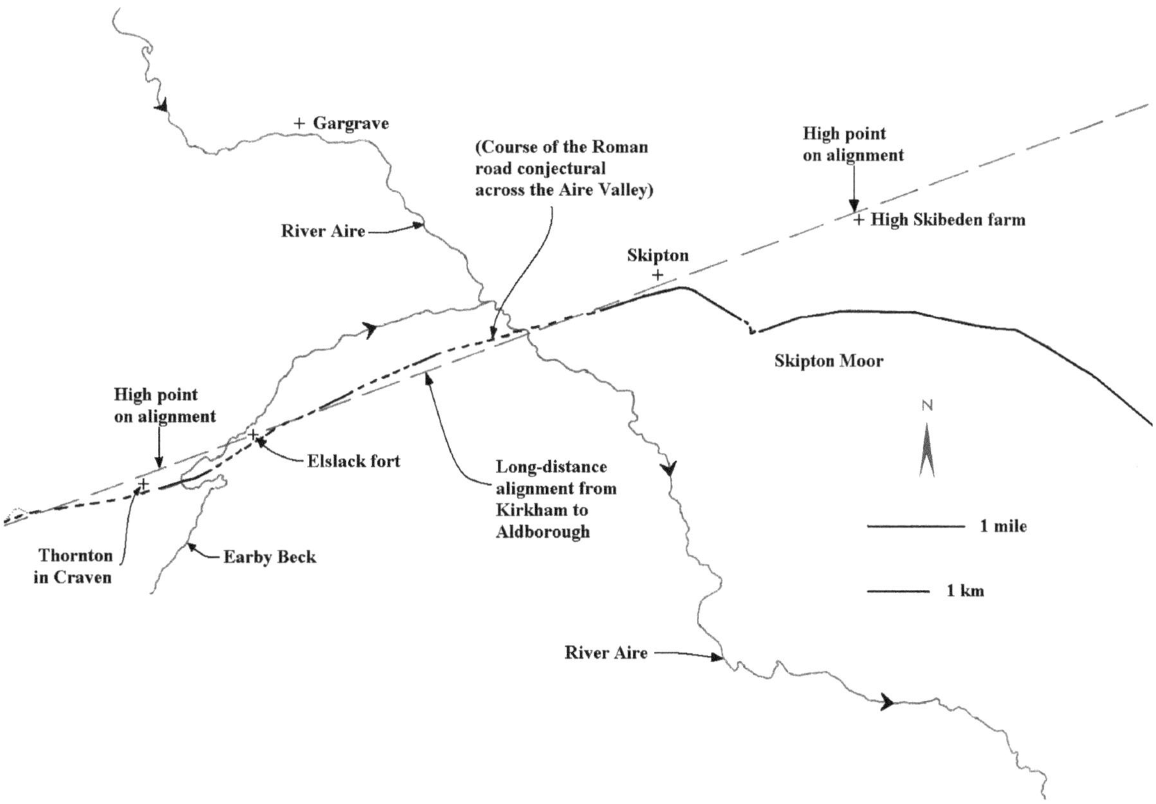

Figure 75 Course of the cross-country Roman road from Thornton in Craven past the fort at Elslack and on to Skipton Moor

[98] Information provided by Mike Haken, *pers. comm.*

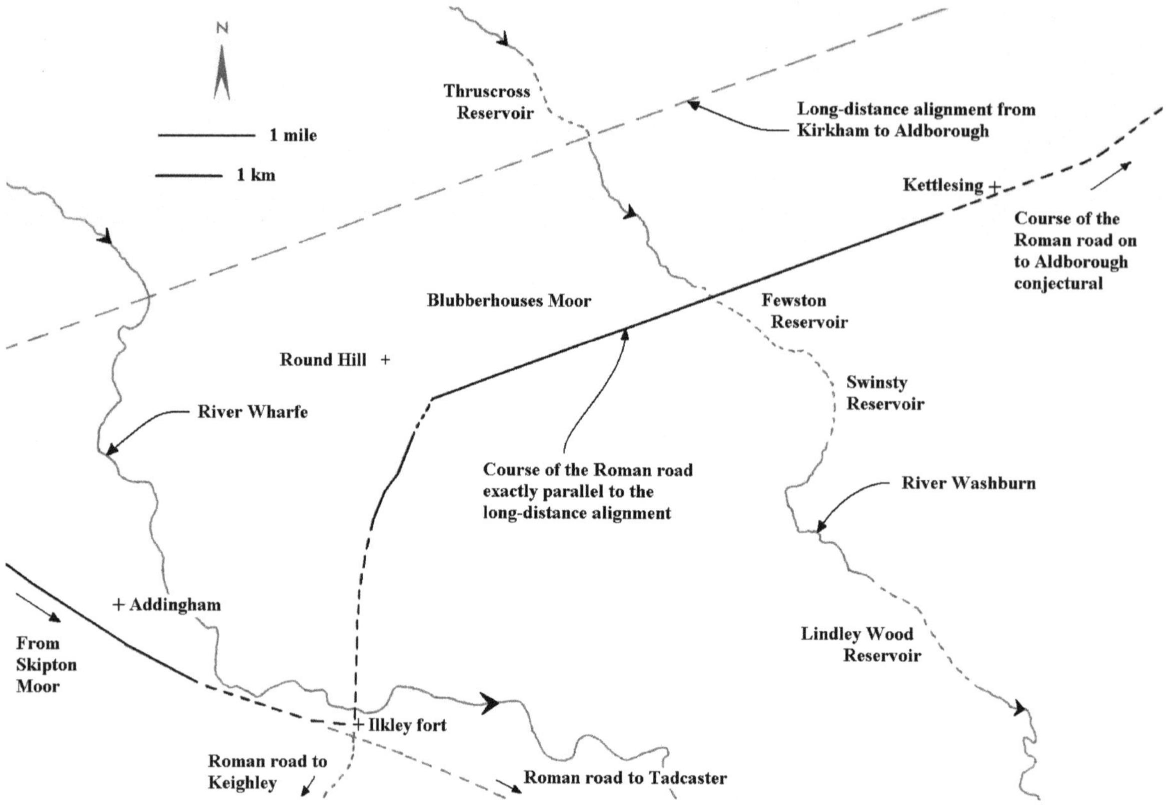

Figure 76 Course of the cross-country Roman road past Addingham to the fort at Ilkley and then returning to run parallel to the long-distance alignment across Blubberhouses Moor

The answer to this may be found at the eastern end of the alignment. As already mentioned, from the excavations conducted by Mike Bishop at Roecliffe beside the River Ure, it appears likely that an original Roman crossing-place of the Ure, dating from the early AD 70s, had been superseded by a new crossing-point over the Ure at around the mid AD 80s.[99] Whether the location of Aldborough had been determined by means of the long-distance alignment through the Craven Gap or as part of a putative centuriation scheme in the Vale of York, it suggests that the Kirkham to Aldborough alignment had been set out at about that time. Although it is understood that Aldborough did not become a civitas capital until Hadrianic times,[100] or possibly even later, it seems inconceivable that such an important crossing-point – carrying Dere Street across the River Ure – would not have been protected by a Roman fort in its earlier days. Although the location of such a fort continues to remain under investigation at this time, the finds from Aldborough do suggest a Roman presence in the pre-Hadrianic period.[101]

This rearrangement of the crossings of the River Ure may have coincided with a phase of consolidation of the network of Roman forts and roads in northern England at around the AD 80s. For instance, it appears likely to have been at this same time that the Romans shifted their military base by the River Tyne from Red House to Corbridge, and planned the course of Dere Street to run through the new site. On balance, therefore, it appears likely that the alignment from Kirkham to Aldborough may have been laid out in the early AD 80s, with the detailed course of the cross-country road on the ground probably planned out at the same time or soon afterwards. As to when the construction of the road or roads took place, there are no indicators of when this might have occurred, although as with Dere Street and possibly the Western Main Road, this may well have been conducted in a piecemeal fashion and over an extended period of time, as and when priorities demanded and resources allowed.

For reference, the earliest periods for the occupation of the forts along this cross-country route are currently considered to be:

Kirkham	On the basis of finds, principally the Samian ware, occupation began around AD 80.[102]
Ribchester	Before Agricola and probably as early as the governorship of Cerialis[103]
Elslack	The fort was established in the governorship of Agricola[104]
Ilkley	The first fort was Agricolan[105]
Aldborough	Likely to have superseded Roecliffe as the crossing-point of the River Ure in the AD 80s[106]

[99] Bishop, 2005, 218
[100] Wacher, 1975, 399
[101] Bishop, 2005, 218 and note 202

[102] Bidwell and Hodgson, 2009, 103
[103] *Ibid*, 99
[104] *Ibid*, 105
[105] *Ibid*, 105
[106] *Ibid*, 139

Finally, for this particular network of Roman roads, it is worth returning to the question of why its planners seemed to have experienced such difficulty in adhering exactly to the long-distance alignment between Downham and Skipton. It is notable that this is a difficulty which appears to have applied all the way along this stretch of road. One possibility, at least, is that although the planners of the road may well have been instructed to use this alignment as a means of setting out a course through the Craven Gap, the alignment itself may not have been ideally marked out for the purpose of laying out the course of the line upon the ground.

Thus, for instance, as indicated above, the alignment might only have been marked out by its surveyors from high point to high point, such as (a) on Longridge Fell, (b) by Great Todber farm, (c) to the east of Thornton in Craven, and (d) beside High Skibeden farm to the east of Skipton. See Figures 73 and 75. The difficulty of sighting, in daylight, upon the latter pair of points against the backdrop of Blubberhouses Moor has already been noted. In addition, although the eastern shoulder of Longridge Fell might have offered a fine vantage-point from which to determine a line though the Craven Gap, it is a less-than-ideal location to be used as a point upon which to sight. Thus, when the silhouette of Longridge Fell is viewed from Great Todber farm, this point stands not upon the highest part of the Fell but part-way down a smoothly-sloping shoulder on its southern side. This could make the point a problem to locate precisely from a distance.

Hence, especially in conditions of imperfect visibility, the possibility of such widely spaced-out markers may go some way to explaining the difficulties that the planners of the Roman road here appear to have experienced, in keeping exactly to what may be presumed to have been the ordained long-distance alignment.

Similarly, with the stretch of the Roman road which runs parallel to the Kirkham to Aldborough alignment across Blubberhouses Moor, it is possible that the alignment adopted for the road may not have been marked out, likewise, in an ideal manner for road-planning purposes. If the alignment had initially been surveyed as part of a fully worked-out scheme of centuriation, then it can be expected that there ought to have been markers spaced at every 776.5 yards (710m) along its length. If, however, the scheme should never have been completed, and the line only marked out in skeleton fashion, possibly from high point to high point, then this might explain the gradual divergence of up to 25 yards (c.23 m) which can be detected in the middle of the road's descent (or ascent) across the moor. Thus, with reference to Figure 76, whilst a marker on the high ground to the east of Fewston Reservoir would have been visible from all the way down the course of the road across Blubberhouses Moor, there are places along the line across the Moor from which any marker to the west, on the shoulder of Round Hill, would not have been visible. Hence if the planners of the Roman road should have been working here from east to west rather than west to east, it is quite possible that they could have strayed a little off-line in the process.

Elsewhere, to the west of Downham Park, the planners of the Roman road seem to have had no difficulty in setting out the course of the road exactly upon their adopted alignments. See Figure 72. Here, however, the alignments may have been of their choice, adopted for sound tactical reasons, and set out by them accordingly for road-planning purposes. Hence their difficulties in adhering to any alignment would have been minimised.

Overall, therefore, it can be seen that this cross-country Roman road or network of roads provides an interesting example of the variety of directives and means by which the planners of Roman roads may have been obliged to operate when setting out the courses of their structures upon the ground.

Section 9: Example of the surveying and planning of the road called the Devil's Causeway from Corbridge to Tweedmouth

The Roman road known as the Devil's Causeway is considered to have branched from Dere Street near Beukley, just north of Hadrian's Wall, and then run through Northumberland to an as-yet undiscovered Roman site at the mouth of the Tweed opposite Berwick-upon-Tweed. Thus, although the exact details of the line at each end of this Roman road remain unclear, it does seem fair to describe the road as running from Corbridge to Tweedmouth.

The course of this Roman road is almost invisible on the ground nowadays, but as reported by Margary[107] it was thoroughly mapped by H. MacLauchlan in 1864 and updated in minor respects following excavations by R. P. Wright some years later. The outcome of these investigations is understood to be the course which is marked on the current Ordnance Survey maps, and reliance had to be placed upon this mapping completely in the current exercise.

The incentive to examine the planning of this road came late in 2011, in response to a question posed by Paul Bidwell[108] about the possibility that a Roman road might have run northwards from Newcastle upon Tyne to Berwick. A number of options for the possible course of such a road were considered, and these will be discussed later in this section. One of the options, though, was that the northern part of the Devil's Causeway might have formed a portion of a road from Newcastle. This was because, from Tweedmouth down to Weldon Bridge, which lies between Rothbury and Morpeth, the general line of the Devil's Causeway does appear to be heading towards Newcastle. Hence the opportunity was taken to examine the planning of the Devil's Causeway throughout.

The planning of the Devil's Causeway – the southern arm

For the Roman planners of the road, a direct line between Corbridge and the mouth of the Tweed would have been blocked by the Cheviot Hills and their southern outliers. See Figure 77. However, rather than rounding the flanks of these hills via a series of short straights, what is striking about the Devil's Causeway is that it accomplishes its passage along just two broadly straight arms. The first arm runs very directly north-eastwards for 14 or more miles (22.5 km) from Ryal, north of Corbridge, to Whinney Hill. Then, after a short linking stretch from Whinney Hill to near Weldon Bridge, the second arm, rather less straight, runs west of north for some 35 miles (c.56 km) from above Weldon Bridge to Tweedmouth, passing the Roman fort at Learchild on the way.

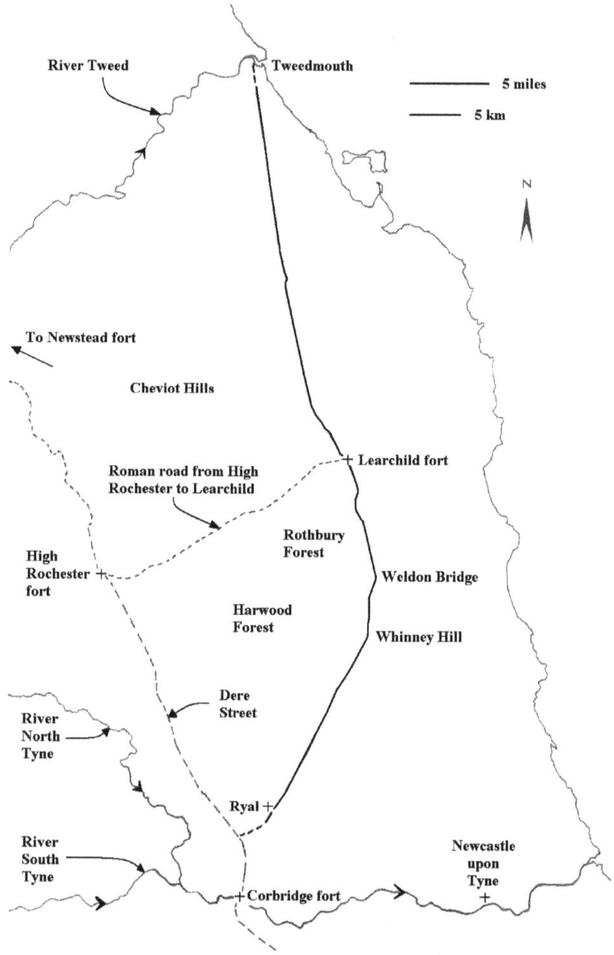

Figure 77 Course of the Devil's Causeway Roman road from north of Corbridge to Tweedmouth opposite Berwick-upon-Tweed

Inspection of the first arm, from Ryal to Whinney Hill, revealed that the final 2 miles (3.2 km) of the Devil's Causeway up to Whinney Hill appeared to be exactly aligned upon the Portgate, where Dere Street passed through Hadrian's Wall. See Figure 78. This discovery came as a complete surprise, and at first it was assumed that this must have been a coincidence. However, just in case it might not have been a coincidence, an examination was undertaken into how the Roman surveyors might have been able to achieve such an alignment. Figure 79 shows the profile of the land between the Portgate and Whinney Hill, and from this it can be seen that there is a single high point near Kearsley farm from which both the Portgate and Whinney Hill could be seen. Thus it would have been simple for the Romans to have set out such an alignment, as follows.

[107] Margary, 1973, 478-80

[108] Paul Bidwell, *pers. comm.*

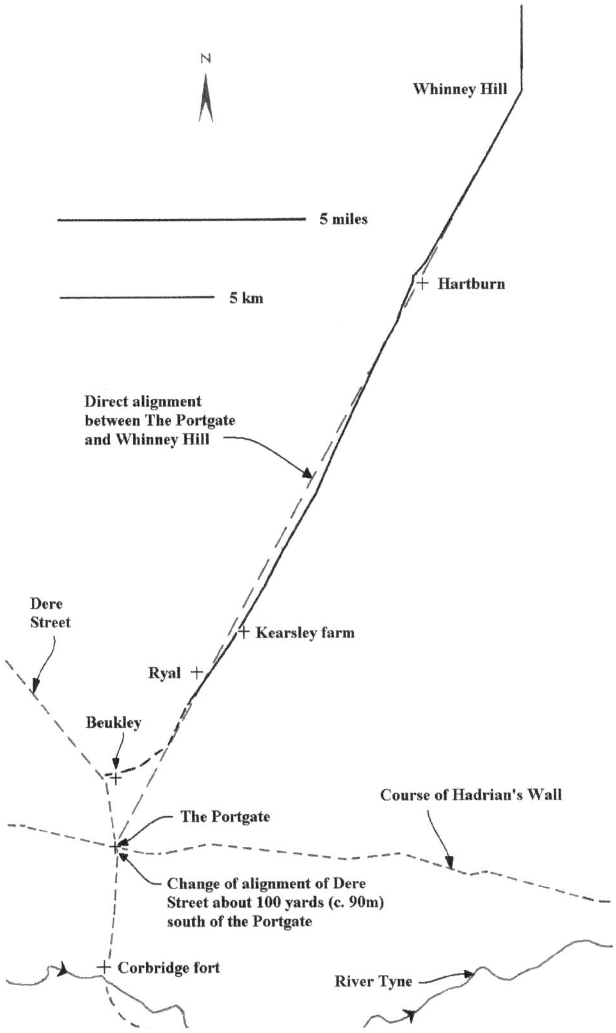

Figure 78 Course of the Devil's Causeway from Beukley to Whinney Hill

Kearsley stands some 12¼ miles (19.6 km) from Whinney Hill, so that although on a clear day Kearsley can certainly be made out from Whinney Hill it would probably have been impossible to have signalled instructions between the two places in daylight. However, by using lights at night, simultaneously on Whinney Hill, at Kearsley, and on the Portgate, it would have been easy for the team at Kearsley to move their light and surveying instrument until they stood exactly in line with the lights at each end.

It should be noted that this process would have enabled the exact bearing of the alignment to have been set at both the Portgate and at Whinney Hill, so that if the Roman road had been laid out from either end point, this would be where it would be most likely to have followed the alignment most closely. As already noted, the Romans appear to have foregone use of the alignment at the Portgate end, commencing construction of their road as a branch from Dere Street near Beukley. However, from Whinney Hill, the fact that the initial alignment appears to have been set out exactly upon the Portgate, coupled with the probability that it would have been practical for the Roman surveyors to have arranged it so, makes it appear unlikely that this alignment was a coincidence.

In response to this conclusion, both Humphrey Welfare[109] and Nick Hodgson[110] have suggested that the alignment could have been set out not from the Portgate but from the point at about NY 988 685 where Dere Street breasts the rise coming up from Corbridge and makes a slight turn to the west of north to head towards Beukley. This point lies about 100 yards (c. 90 m) to the south of the Portgate, and an alignment upon it would have been less than ½° different from the alignment to the Portgate.

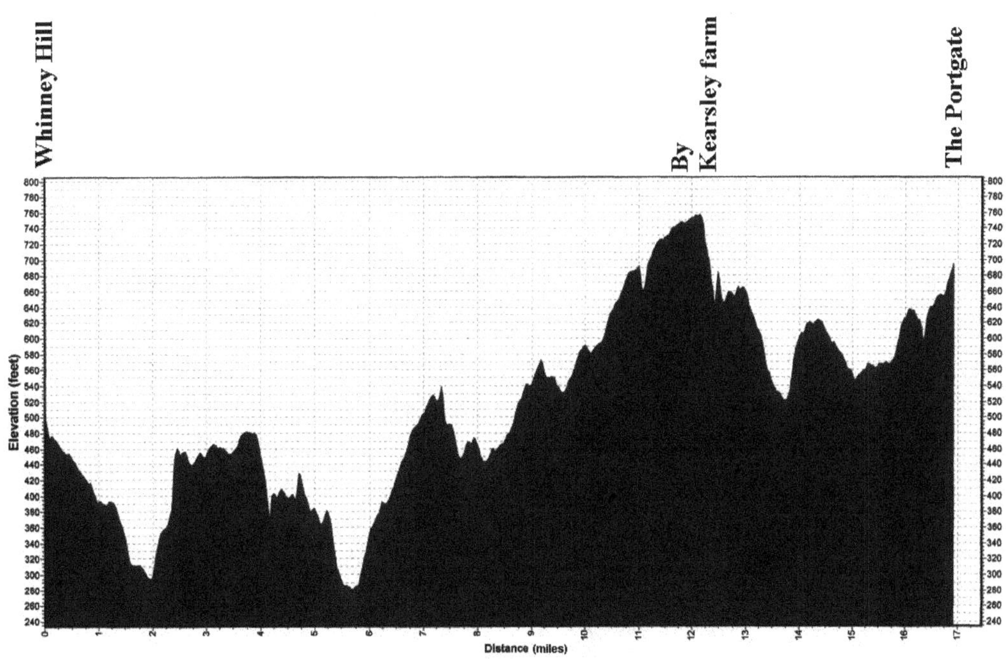

Figure 79 Profile of the landscape on a direct alignment between Whinney Hill and the Portgate

[109] Humphrey Welfare, *pers. comm.*
[110] Nick Hodgson, *pers. comm.*

Trying to differentiate between two possible alignments as close as this by examining the lines on a map is scarcely practical. It would have to depend upon the combined precision of modern mapping, MacLauchlan's surveying and Roman road-making. Instead, the possible design intentions may be considered. Firstly, if the alignment had been set out from south-west to north-east, there are no physical features today, either by Kearsley farm or on Whinney Hill, which might have prompted the Roman surveyors to choose such a direction. It therefore appears more likely that the Roman surveyors were working from Whinney Hill and aiming for a target – either the Portgate or else the brow at which Dere Street happened to turn on its way northwards. Both are possible, but to the author, the Portgate would seem to have been the more perceptible target. Besides, if the Romans had simply been intending to make a junction with Dere Street, they might have done better to have aimed for the point where Dere Street turns near Beukley.

The implications which these alternative possibilities have for the date of the Devil's Causeway are discussed later in this section. In the meantime, the laying-out of this first arm of the Devil's Causeway will now be examined in more detail.

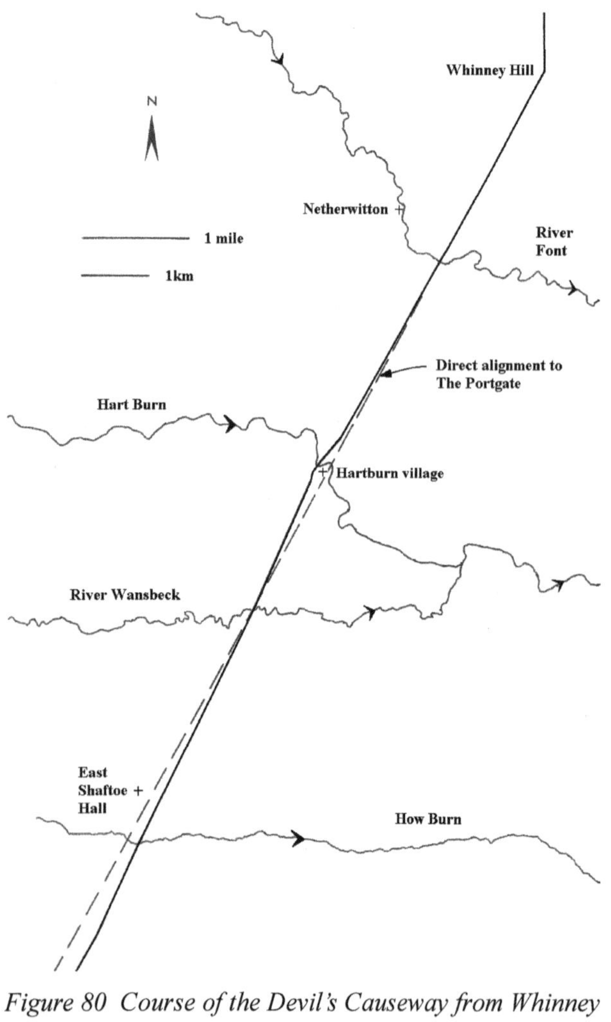

Figure 80 Course of the Devil's Causeway from Whinney Hill past Hartburn village and the River Wansbeck to the How Burn

Starting at Whinney Hill, it can be seen on Figure 80 that the long-distance alignment is followed exactly for only about two miles (3.2 km), to the south-east of Netherwitton. After this a change of direction occurs on the north side of the River Font, leading to a deviation to cross the Hart Burn. Apart from the area to the south of Ryal, where the road breaks away from the alignment altogether in order to join Dere Street, this deviation appears to be the only one which was required (as distinct from taken) along this arm of the Roman road. It appears to have been necessary in order to cross the Hart Burn at a suitable point above the village of Hartburn, through which the burn flows in quite a canyon in places. The return to the long-distance alignment was made at the crossing of the River Wansbeck.

From this point south-westwards, it is notable that the road did not run along the alignment exactly. This may be because the long-distance alignment had not been set out closely on the ground once the bearing between the two end-points had been established. For practical reasons, this may have been judged unnecessary, because the landscape between Whinney Hill and the south-western end of the alignment is quite benign, so that straying a short distance from the alignment would have made no difference to the ease or speed of constructing the road, nor to its overall directness. In the event, between Ryal and Whinney Hill, the road as built never runs more than 330 yards (c.302 m) from the long-distance alignment.

The first instance of this possible lack of marking-out occurs at the River Wansbeck, where, after the return from the deviation, a slight change of direction was made to the road, but not sufficiently to resume its correspondence with the long-distance alignment. At the Wansbeck crossing, neither of the two end-points of the alignment would have been visible, and if no other markers had been present, then the planner laying out the course of the road would have had to guess by how much the direction of the road would have to be changed to restore its allegiance to the alignment, and may thus have slightly misjudged it. As can be seen on Figure 80, the result was a continuation of the Roman road on a somewhat less divergent course on the opposite side of the long-distance alignment. This new line was then pursued past East Shaftoe Hall to the crossing of the How Burn, where it seems that a turn was made again on the northern bank.

Up to this point, the direction of setting-out appears to have been south-westwards from Whinney Hill. From the southern end of the road, however, the setting-out appears to have been to the north-east. This can be seen on Figures 81 and 82, where most of the slight changes of direction north of Ryal and up to the How Burn appear to lie on north-facing slopes. If so, this would have meant that the planners setting out the southern end of the Roman road, having branched from Dere Street, would have had to work its line across the countryside to pick up the long-distance alignment. They appear to have achieved this before Ryal, but then overshot it to run some 220 yards (c.200 m) to the east of the high point by Kearsley. It may

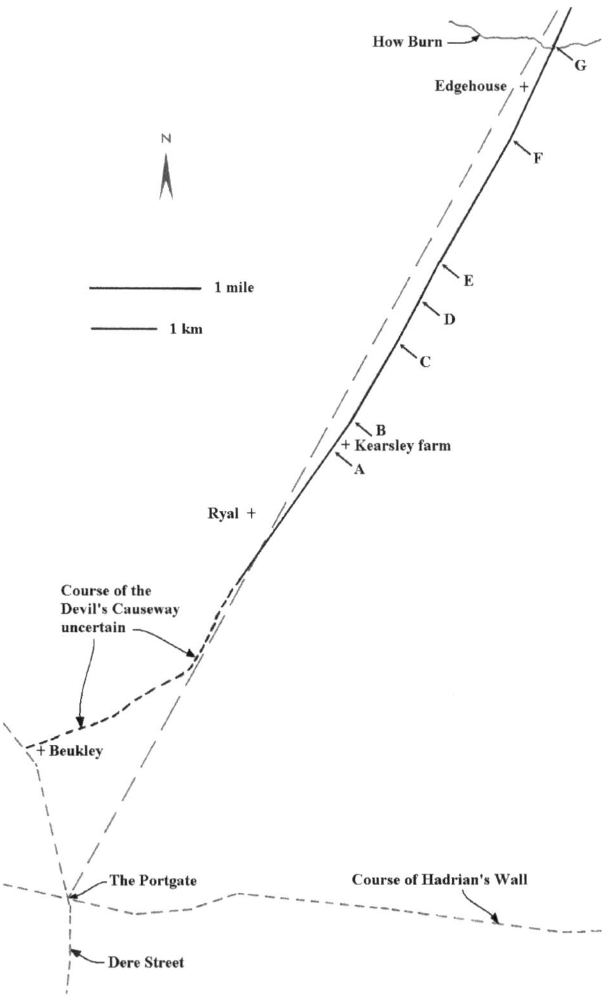

Figure 81 Course of the Devil's Causeway from Beukley to the How Burn, showing the slight changes of alignment along the way

have been that any marker there had been difficult to see from near Ryal, or else that the Roman planners may have preferred to make their own sightings – in daylight – a little to the east of the high point, which is most likely to have been established previously by their surveyors at night. Certainly, beyond Kearsley, the road as built proceeds in a series of short straights which look like successive but not entirely successful attempts to sight upon a marker for the long-distance alignment on Whinney Hill, some 10 to 12 miles (16 to 19 km) away.

The junction between the two settings-out would appear to have occurred between the How Burn and a point to the south of Edgehouse, at about NZ 053 797 (point F on Figure 81). It appears likely that this clinching alignment would have been set out by the team working from the north-east to the south-west. This is because of the turn on the northern bank of the How Burn (point G on Figure 81), and the fact that, as can be seen on Figure 82, the northern team would have been sighting uphill to the skyline towards point F.

The planning of the Devil's Causeway – the northern arm

If the long-distance alignment of the southern arm of the Devil's Causeway had been a surprise, that of the northern arm proved to be even more so. The final 11 miles (17.7 km) of this Roman road run very straight, from NU 023 335 near Hetton Law farm to the final point at which the road is mapped before Tweedmouth, at NU 998 507, near Springhill farm. Remarkably, this stretch appears to be

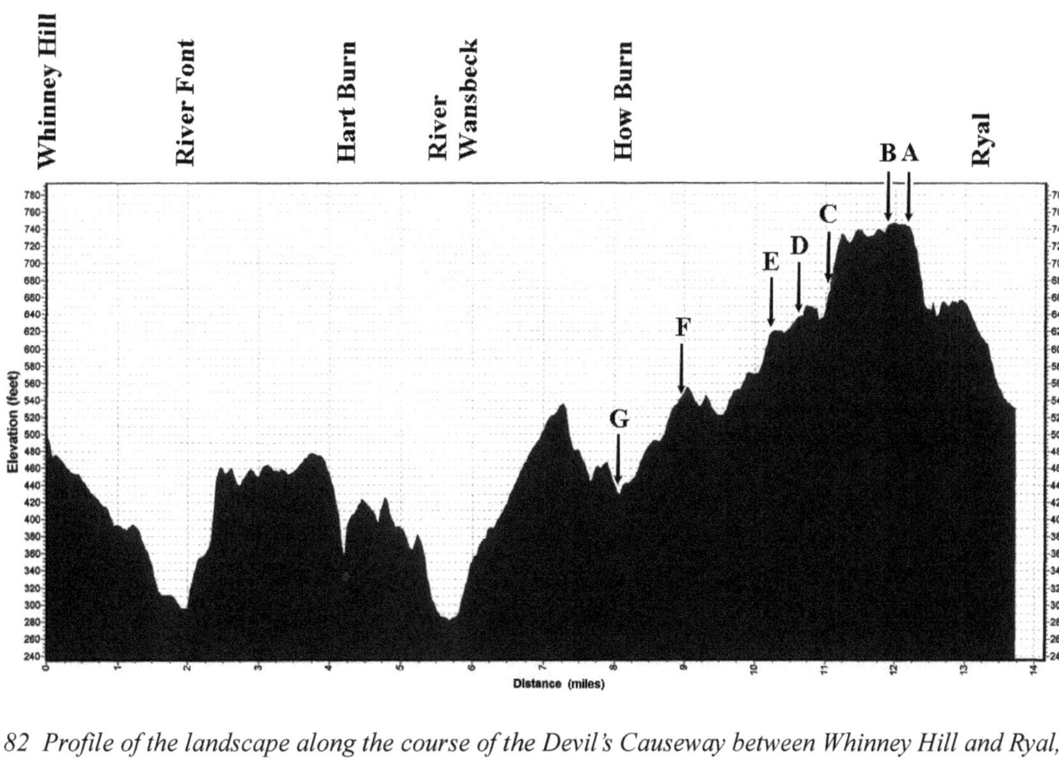

Figure 82 Profile of the landscape along the course of the Devil's Causeway between Whinney Hill and Ryal, showing the locations of the slight changes of alignment indicated on Figure 81

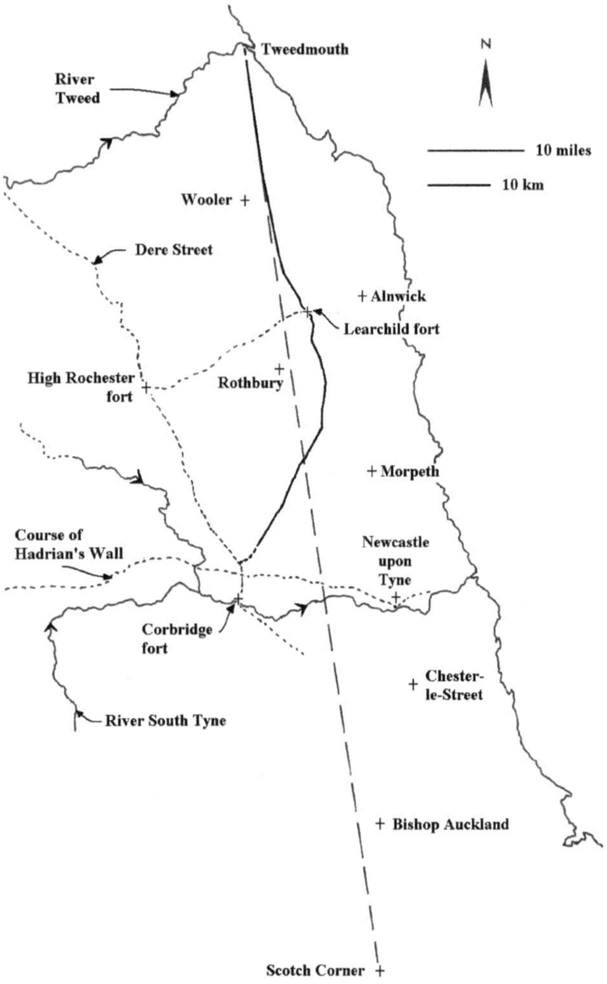

Figure 83 Course of the Devil's Causeway, showing the alignment of its northernmost length with Scotch Corner

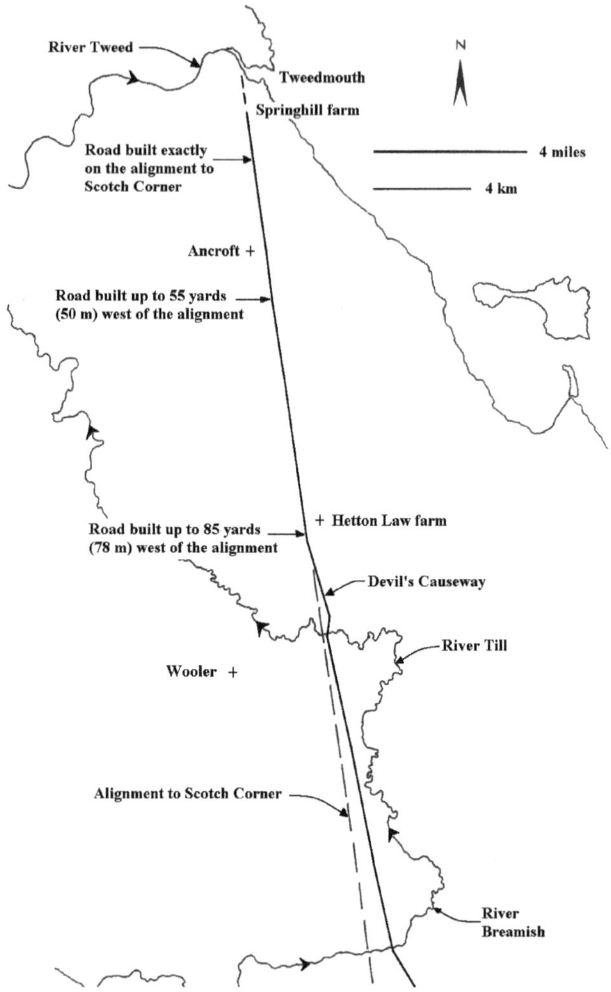

Figure 84 Course of the northernmost stretch of the Devil's Causeway, showing the detail of its relationship with the alignment to Scotch Corner

aligned upon Scotch Corner, more than 90 miles (145 km) away. In fact, the final 3½ miles (5.6 km), from NU 007 453 (Ancroft) to near Springhill farm appears to be exactly aligned upon the Roman road junction at Scotch Corner. See Figures 83 and 84.

Again, the initial reaction was that this had to be a coincidence. This was because it could not be conceived how Roman surveyors would have been able to set out such an alignment with such precision over such a distance. Quite apart from the distances involved, there is no single high point between the mouth of the Tweed and Scotch Corner from which both ends of such an alignment would have been visible. Michael Lewis has proposed a method using geometrical construction that could have been employed by the Romans to determine an exact bearing between two points that were not intervisible,[111] but with multiple intermediate high points it seems that this could have been a very difficult and laborious process to have attempted in this case. The remaining possibility is that the Romans had, by the time that the Devil's Causeway came to be set out, extended at least the skeleton of a scheme of centuriation up to the mouth of the Tweed, if not beyond. John Peterson is ready to accept the possibility that part of the English Lake District had been covered by a scheme of centuriation,[112] so possibly the existence of one even further north on the other side of the country may not be so far-fetched.

Despite these reservations about how such a feat of surveying could have been accomplished, the conviction grew that this alignment was unlikely to have happened by chance. The exactness of the coincidence of the alignments, especially over the final 3½ miles (5.6 km), is especially striking. Moreover, the way that the line of the Devil's Causeway, as built, comes on to the line from Scotch Corner near Hetton Law farm is very reminiscent of the way in which Dere Street, near Dishforth, comes

[111] Lewis, 2001, 232-3

[112] Peterson, 2006, Item No. 1

on to the long-distance alignment from Tadcaster (see Figure 23), and the way in which the Western Main Roman Road, after climbing up Croasdale and then descending the Tatham Fells, comes on to the alignment from Jeffry Hill to Casterton in Lunesdale (see Figure 60). In addition, the lie of the land southwards from near Springhill farm to near Hetton Law farm makes it an almost ideal choice for the course of a road to the south, and this is surely something upon which the surveyor of an original long-distance alignment would have reported back, when debriefing upon the completion of their assignment.

For the Devil's Causeway to have continued southwards along such an alignment would, however, have brought it eventually to the precipitous scenery around Rothbury. Hence a divergence from the alignment would have been required, and this was taken at the southern end of the plateau, near Hetton Law farm, where a new alignment was prefaced by a short deviation to cross the River Till near Heatheryhall. See Figure 85. The new alignment does not quite line up with where the Devil's Causeway turns to begin the descent to the River Till, at NU 023 335. It might have been set out from a point about ½ mile (0.8 km) to the south. Either way, both of these points are on south-facing slopes, suggesting that the Devil's Causeway was being set out from north to south here, as also appears to have been the case on Whinney Hill, mentioned above. From this point southwards, it should be noted, the alignment to Scotch Corner appears to have played no further part in determining the course of this Roman road.

Having crossed the River Till and taken up the new alignment, the Devil's Causeway then ran to the west of the river for some distance, adopting a surprisingly low-lying line beside it in places. Along the way the road passed by a possible Roman site at NY 049 209, near Wooperton, where Roman finds have recently been reported. Overall, between the Till and the crossing of the River Coquet above Weldon Bridge, the countryside which the Roman road traversed largely consists of a gently undulating landscape but it is one which is punctuated by two not insignificant rivers – the Breamish and the Aln – and some areas of low-lying ground that might have been swampy in Roman times. The Roman fort at Learchild seems to have been located near the perimeter of one such basin. In these conditions, the course of any road is likely to have been determined by the need to avoid the low-lying ground and to choose the most suitable crossing-places over the rivers. It is notable that not only did the Devil's Causeway and the A697 road adopt adjacent crossing points over the River Aln and the River Breamish, but so did the accompanying railway from Alnwick to Wooler, whilst between these points all three of these linear constructions tended to run some distance apart.

Across this gently undulating countryside, the only significant rise in altitude occurs just to the south of the Roman fort at Learchild, where the Roman road had to run over a ridge of hills which extends from Rothbury to Alnwick. These hills are not particularly steep, but it

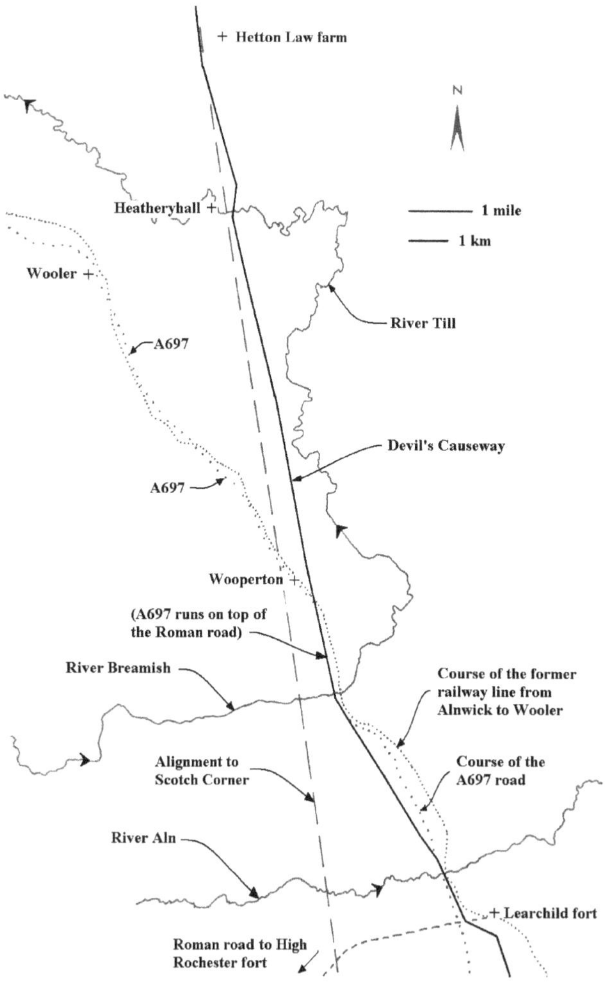

Figure 85 Course of the Devil's Causeway from the River Till to the fort at Learchild

can be imagined that the aim of any road builder would have been to keep their traverse of the ridge as short and as low-lying as possible in order to minimise the risk of blockage by snow in the winter. Perhaps characteristically, the Roman road took a more direct route over this ridge than the A697 road, but it is one which involved a double climb, with a steep rise up from the fort to a summit near Desmesne, then a descent to the Edlingham Burn before climbing again to Longframlington Common. See Figure 86. In contrast, the A697 road makes a superbly graded ascent of the western side of the hill opposite Desmesne and then completes a traverse of the ridge in the shortest possible distance before descending to meet up with the Roman road near Framlington Gate. If the aim had been to minimise the distance spent at height, then the course here of the A697 makes the planning of the Roman road look like something of a misjudgement, although of course we cannot entirely tell what issues may have faced the Roman surveyors at the time. What can be said is that the rather awkward course of the Devil's Causeway near the fort at Learchild, together with the general alignments adopted between the River Till and the River Coquet, make it fairly clear that at least this part of the Roman road seems to have been laid out expressly to service the fort at Learchild.

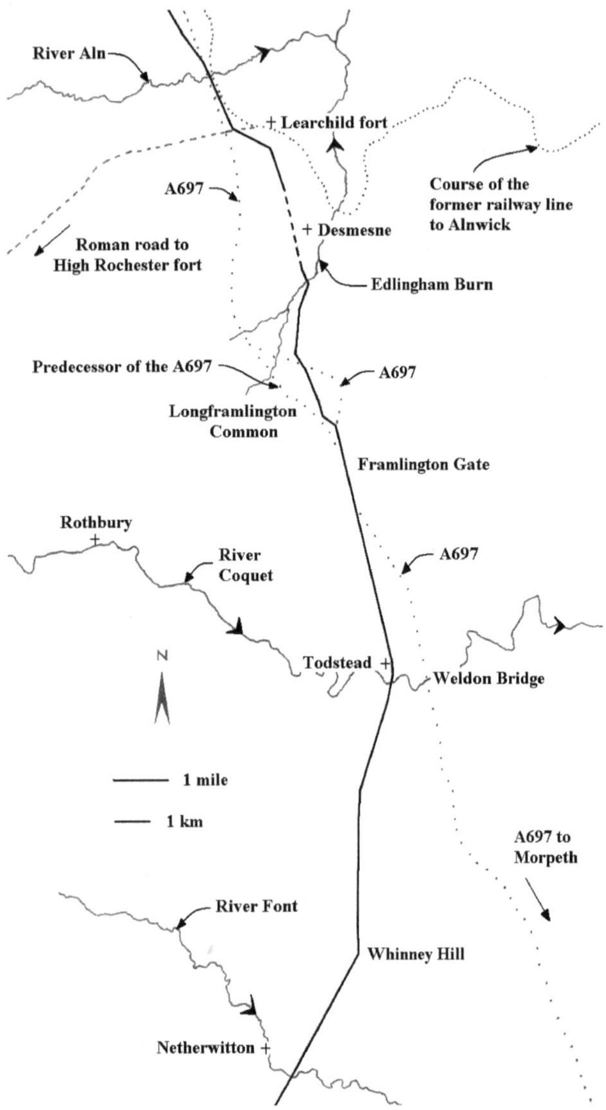

Figure 86 Course of the Devil's Causeway from the fort at Learchild to Whinney Hill

After crossing the Coquet, near Weldon Bridge, the Devil's Causeway and the A697 road part company decisively, with the Roman road heading south-westwards and then south towards Whinney Hill whilst the A697 heads directly on for Morpeth. This short dog-leg of the Roman road, from near Weldon Bridge to Whinney Hill, can simply be seen as a means of attaining a convenient south-facing viewpoint from which to set out a direct line to the south-west. As it happens, it is difficult to diagnose the direction in which this part of the Roman road might have been planned. The impression gained is that the sighting-point to the south-west on Whinney Hill might have been chosen first, and then the line of the dog-leg simply set out northwards until the line of the road coming down from Longframlington Common could be seen above the crossing of the River Coquet, near Todstead. Certainly the laying-out of this short link between the two arms of the Devil's Causeway seems to be separate from the planning of either limb.

Discussion about the date of the Devil's Causeway

If the south-western end of the Devil's Causeway branched off Dere Street, perhaps the firmest point which can be made about the date of construction of the Devil's Causeway is that it would have come after Dere Street had been built. A general assessment of the date of the Devil's Causeway has been well summarised by Tim Gates and Richard Hewitt[113] and recently developed further by Nick Hodgson.[114] Based upon the Roman finds from Learchild and near Wooperton, the lack of stone buildings and inscriptions from the fort at Learchild and no sign of any stone buildings or other Roman remains from the mouth of the Tweed, the received opinion at present is that the Devil's Causeway was of pre-Hadrianic date and that it probably went out of use when Hadrian's Wall was established.

Hence to find that the long-distance alignment for the southern arm of the Devil's Causeway may have been aligned upon the Portgate is certainly contentious. This is because the site of the Portgate would not have come into existence until the line of Hadrian's Wall was being set out. Recently, Erik Graafstal has even suggested that the course of Dere Street near the Portgate may have been realigned when Hadrian's Wall was set out.[115] If so, the alignment of Dere Street from Beukley to NY 988 685, where Dere Street rises over the brow to the south of the Portgate, could even be Hadrianic. However, Erik Graafstal's suggestion was based upon a perception that the Portgate was positioned at exactly 5 Roman miles from where Hadrian's Wall crossed the River North Tyne, and this now appears to be slightly incorrect. Hence it seems more likely that the Portgate is where it is because that was where Dere Street was running at the time when Hadrian's Wall was being set out.

It has to be agreed that it is difficult to place great weight upon the evidence of a single alignment extrapolated over a distance of more than 15 miles (24 km), but perhaps this opens up the possibility that when Hadrian's Wall was being planned, there had been an intention to retain Learchild in commission as an outpost fort north of the Wall, in the way that the forts at Birrens, Netherby, and Bewcastle came to be stationed in advance of the Wall on the western side of the country. As yet there is no evidence for the existence of a fort at Newcastle when Hadrian's Wall was being laid out. Hence the Roman fort at Corbridge and a Roman site at the mouth of the Tweed would have

[113] Gates and Hewitt, 2007, 25-9

[114] Hodgson, 2012, 211-3. In his discussion of the evidence, Nick Hodgson advances the attractive idea that the Devil's Causeway might have been conceived as the continuation of the Roman frontier to the east of Corbridge. Thus rather than extending the Stanegate eastwards from Corbridge towards the mouth of the Tyne (for which no reliable evidence has yet been found), the Roman commanders may have sought instead to continue their frontier north-eastwards beyond Corbridge to reach the North Sea at the mouth of the Tweed. This interpretation could place the date of the construction of the Devil's Causeway to around AD 105 or later, which is considered to be the likeliest date for the construction of the Stanegate. See Poulter, 1998.

[115] Graafstal, 2012, 143

been the places from which Learchild would have been most naturally serviced, and at that time it may have been considered worthwhile to lay out a road connecting Learchild with these two sites. Thereafter, a change of plan – a not uncommon feature during the construction of Hadrian's Wall – could have led to the abandonment of both Learchild and the Devil's Causeway not long after the Hadrianic period had commenced. That change of plan could, for instance, have included the decision to locate forts along the line of Hadrian's Wall, including to the east of Corbridge, and this in turn might have been considered sufficient to make Learchild and the Devil's Causeway redundant.

The possibility of a Roman road running northwards from Newcastle upon Tyne

As explained at the beginning of this section, the study of the planning of the Devil's Causeway arose almost as a by-product of a more general question. This was about the possible existence of a Roman road running northwards from Newcastle to Berwick-upon-Tweed, and this question will now be considered.

To begin with, although it is agreed that there was a Roman bridge over the River Tyne at Newcastle, it is not certain that it or its associated fort were constructed before the end of the second or the beginning of the third centuries.[116] Nor is it known for certain where the bridge stood. It is often thought to have been where the medieval bridge crossed the Tyne, which is understood to be where the swing bridge does now, but until firm evidence of the location of the Roman bridge can be uncovered this can be no more than an assumption, albeit a probable one.[117]

In view of the number of battles which have been recorded along or close by it since Roman times, the prime candidate for the course of a Roman road running northwards from Newcastle may appear to be the line of what is now known as the A1 road. John Ogilby's map of 1675 shows that this road was already fully-formed by that date, rather than being just a loose assemblage of routeways over tracts of land.[118] Ogilby's map also shows that the course of the road in 1675 was very largely the same as that of the modern A1 road prior to its recent improvements and realignments for motor traffic. This is shown in Figure 87, in comparison with the line of the Devil's Causeway.

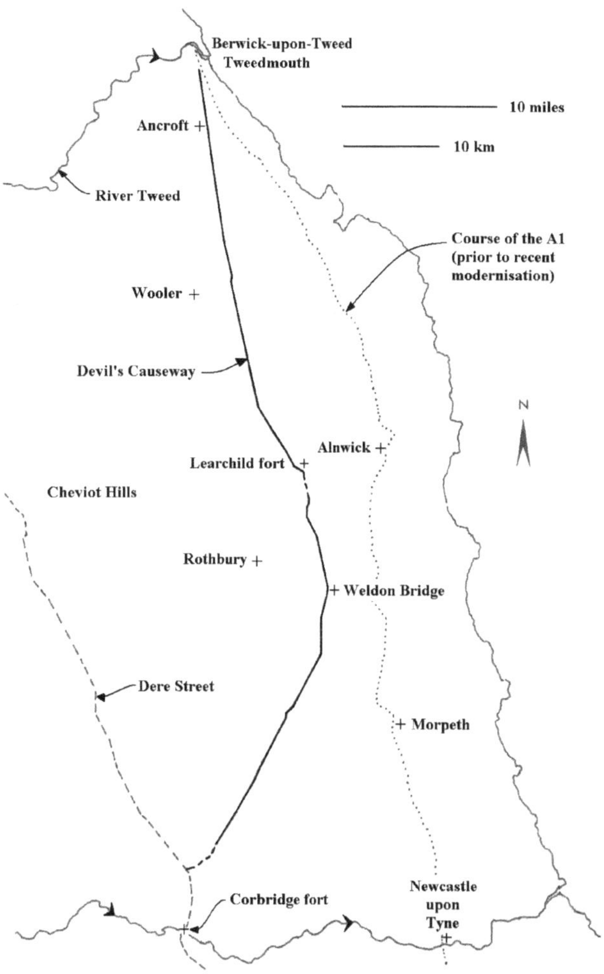

Figure 87 Course of the A1 road (prior to recent modernisation) from Newcastle to Berwick-upon-Tweed, in comparison with the line of the Devil's Causeway

Examination of the unmodernised route of the A1 showed promise at first, when it was observed that the course of the Great North Road towards Gosforth, just north of Newcastle, appeared to be exactly aligned upon the northern end of the medieval bridge over the Tyne. See Figure 88. However, as indicated above, it is not entirely certain that this would have been where the Roman bridgehead would have stood. Moreover, from this point northwards all the way to the mouth of the River Tweed, the course of the former A1 appears to exhibit no sign of Roman alignments or planning. It might be suggested that the unaligned course of the A1 north of Newcastle could be a continuation of the rather rambling nature of the Roman road coming up to Newcastle from the south (Margary's road number 80[119]), but even this road displays some indications of Roman alignments in places, such as between Sedgefield and Coxhoe, and to the north of Chester-le-Street.

[116] Breeze, 2006, 144-5. Paul Bidwell, however, (*pers. comm.*) considers that, on practical grounds, a bridge would have been required at Newcastle from at least the beginning of Hadrian's Wall, and possibly even earlier. In addition, he finds the association of the name of the bridge (*Pons Aelii*) with Hadrian to be compelling. *Aelius* was the family name of Hadrian.

[117] Breeze, 2006, 144-5. Paul Bidwell comments though (*pers. comm.*) that finds of altars and inscriptions, and the course of the Roman road coming up to Newcastle from Chester-le-Street all point to the vicinity of the swing bridge – as does the position of the Roman settlement at Bottle Bank in Gateshead, which overlooks the swing bridge.

[118] Ogilby, 1675. Paul Bidwell adds (*pers. comm.*) that there is documentary evidence from the mid 13th century for the existence of the southern part of this, "the great royal road".

[119] Margary, 1973, 431-3, 441

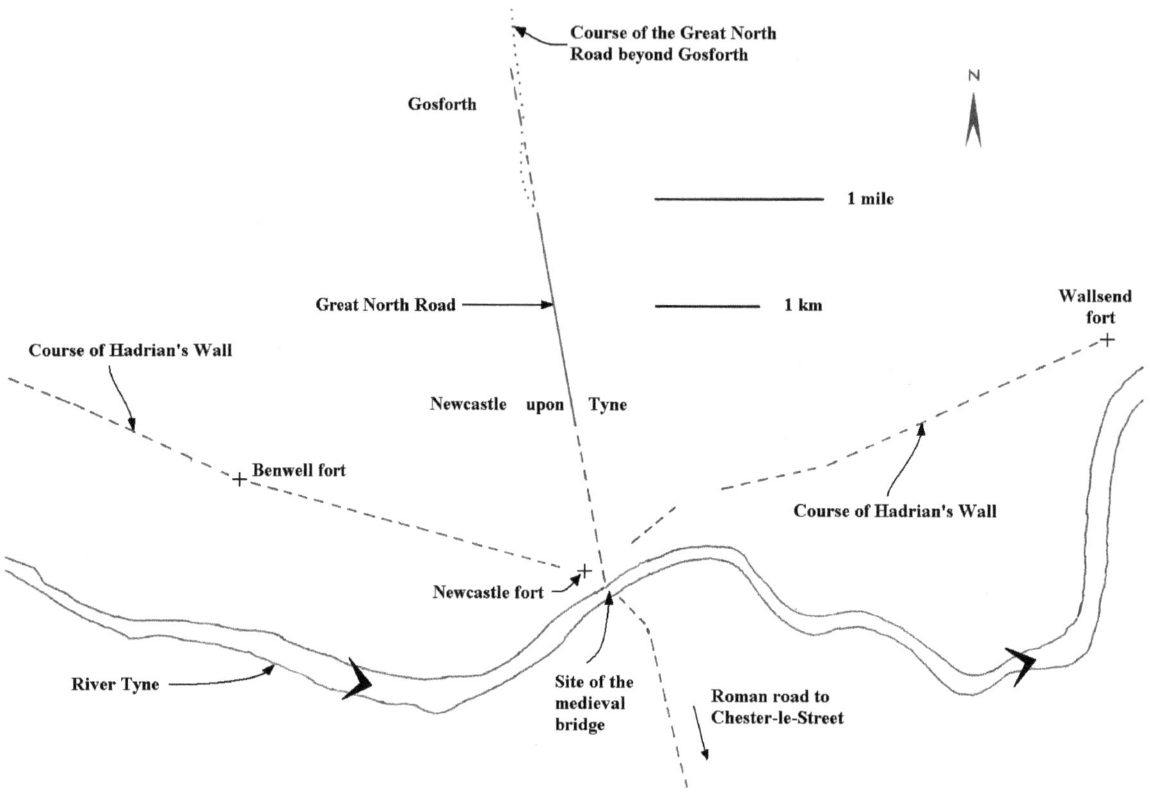

Figure 88 Course of the Great North Road towards Gosforth showing its exact alignment upon the site of the medieval bridge over the Tyne at Newcastle

It is at least theoretically possible that a Roman road to the north of Newcastle might not have been laid out until Antonine times, or even later. By that time the Roman Empire had largely ceased to expand, so that it is also possible that by then there could have been less of a surfeit of cheap labour. This might in turn have caused the planners of any new Roman roads to have given more consideration to economies of construction and operation, thus making their roads less easy to distinguish from roads constructed in more modern times. However, this is all highly speculative and it is difficult to place much weight upon this as a possibility.

Besides, with regard to the former A1 road, there is the further point that as it approaches the mouth of the Tweed it gradually converges upon the line of the Devil's Causeway but appears to acknowledge no recognition of the proximity. It seems to be unlikely that the Romans would have constructed two roads from the south that were aligned so closely together at this point. See Figure 87 again. If the Romans had planned the A1 it can be expected that somewhere north of Ancroft there would have been either a link to the Devil's Causeway or vice versa, and a single route thereafter to Tweedmouth. As it is, there is no indication of such a link. Hence it is considered that if there had been a Roman road between Newcastle and the mouth of the Tweed, its northern portion at least would have been that of the road which is attested – the Devil's Causeway.

As already indicated, the upper arm of the Devil's Causeway, down to near Weldon Bridge, does appear to be heading towards Newcastle. This leads to a number of possibilities for the development of a Roman road network between Tweedmouth, Corbridge, and Newcastle. For instance a road could have been created between Newcastle and Tweedmouth, and then the lower arm of the Devil's Causeway built as a branch from it to Corbridge. Alternatively, the Devil's Causeway could have been built in its entirety first, and then a branch extended from it over the 23 or so miles (37 km) to Newcastle, or, as another alternative, all three arms could have been set out as the interconnecting links of a single concerted plan.

Certainly the two arms of the known course of the Devil's Causeway would appear to have been set out by the same team. The style of planning of both arms of the Devil's Causeway is broadly the same, i.e. long-distance alignments are employed with relatively short deviations to attain local objectives or avoid individual obstacles. This is in marked contrast to the layout of the other Roman road which is known in the area – the road connecting the fort at High Rochester with that at Learchild, which is shown on Figure 83. This road proceeds in a series of short lengths that look as if they were set out on a line-of-sight basis – although it has to be said that the author has not verified this on the ground.

Between Wooperton and Weldon Bridge the course of the Devil's Causeway runs across much the same countryside as the A697 road. However, as already noted, across this same landscape the A697 only touches upon the Devil's Causeway briefly and there appears to be little indication that its builders were seeking to make use of the line of the earlier Roman road. In addition, at its southern end, the A697 runs to Morpeth rather than Newcastle and its course is as devoid of long-distance alignments as the A1. Hence, from all the foregoing, it is considered that if any Roman road were to have been constructed directly from near Weldon Bridge to Newcastle, it appears likely that it would have run somewhat to the west of the A697 rather than underneath it. See Figure 89.

Unfortunately, such a road is likely to be difficult to trace. Not only is the course of the Devil's Causeway almost always invisible on the ground but it can be seen from the air that it is also largely ignored by present-day roads and field boundaries. In addition, perhaps not unexpectedly, it can be seen that the Devil's Causeway is ignored by all significant modern settlements, such as Rothbury, Alnwick, and Wooler. Thus even though Henry MacLauchlan was able to map the course of the Devil's Causeway in the 1860s, it appears that its line had long been abandoned and forgotten by the time that the roads and land divisions and settlements that exist in these parts of Northumberland today were laid out. The impression given, therefore, is that the Devil's Causeway had been abandoned well before enduring post-Roman settlement occurred in this area, and possibly, as already suggested, well before the Roman occupation of Britain had ended.

If so, it can be expected that similar invisibility would apply to the course of any other Roman road which might exist in the area, such as a possible road from near Weldon Bridge to Newcastle. The upshot of all this is that if such a road had ever been built – and one final possibility is that such a Roman road might have been planned but never built – then it appears most likely that it would have run to the west of the A697 but is going to be difficult to find.

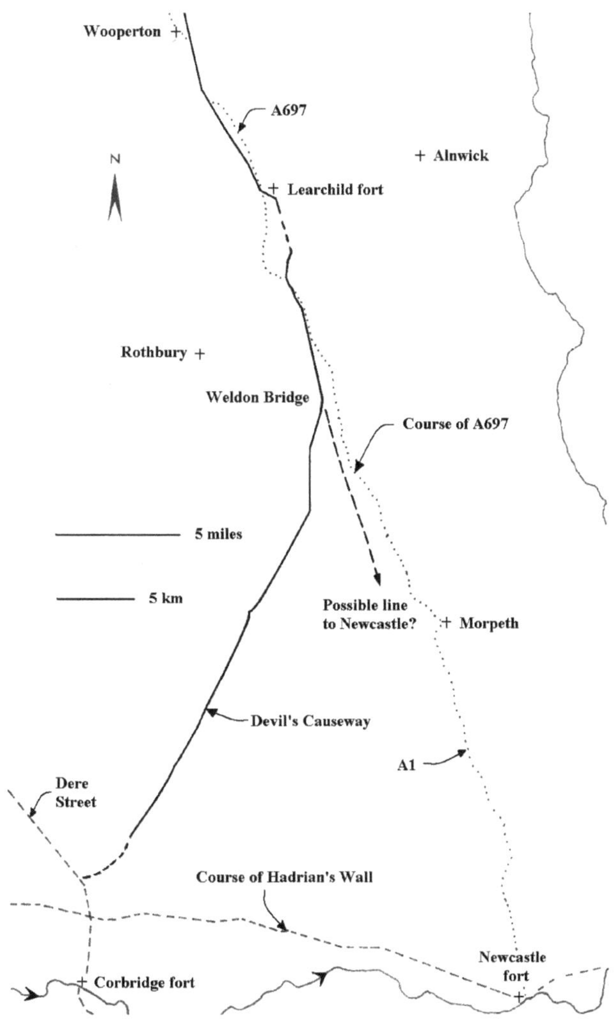

Figure 89 Courses of the Devil's Causeway and the A697, showing the likeliest direction of any Roman road that might have existed from near Weldon Bridge to Newcastle

Section 10: Review of the points emerging from the foregoing analyses and examples

As a prelude to reflecting upon the outcomes of the research reported in this monograph, it may be helpful to revisit the entire process of enquiry and analysis which the author has followed, in his examination of the planning of Roman roads in Northern Britain. Thus this brings together the work reported in BAR 492 as well as in this latest volume.

Starting with a simple question about the direction of planning of a short stretch of Dere Street, the author developed a methodology for diagnosing the direction of planning of Roman roads. This methodology embodied suppositions about how Roman surveyors or engineers, when engaged in setting out the lines of their roads on the ground, would have behaved in certain situations, chiefly when seeking to change direction.

This methodology was tested out along a substantial length of Roman Dere Street from the Vale of York to Newstead in Scotland, and to a very large extent, the results of this fieldwork appeared to confirm that the suppositions about the behaviour of Roman planners had been correct.

However, when the results of these observations were consolidated along the entire course of Dere Street, the patterns in which the teams of Roman surveyors seemed to have been working when setting out the various stretches of their road appeared to have been inexplicably disorganised. This seemed scarcely credible, and various models were therefore applied to this accumulated evidence to try to interpret what processes the planners of the Roman road might have been following when setting out its line.

After considerable investigation, only one model emerged which fitted the evidence all the way along the road. This was that the Roman surveyors had indulged (in the main) in a two-stage setting-out process, in which long-distance alignments had been set out first, and then deviations had been set out from these alignments to meet local objectives.

It was at this point, it should be noted, that the author had moved beyond his original question. At the outset, there had been no intention of attempting to examine the processes by which Roman roads had been planned, beyond diagnosing the directions in which the planners of the roads might have been working. In particular, there had been no inkling that the Romans might have adopted a two-stage planning process, nor that the courses of their roads could be underpinned by long-distance alignments of 20, 30 or even more miles in length. These, therefore, were entirely unexpected discoveries, and as far as the author is aware neither of them had been proposed previously with adequate supporting evidence.[120]

At the suggestion of David Breeze, the author then applied the same test to the course of Hadrian's Wall, and found that its laying-out had been a two-stage process too. In contrast, when the author examined the setting-out of the Antonine Wall in Scotland, it was found that the line of that structure had been created by an entirely different process. Although interesting in itself, this also indicated that the author had not been imagining the two-stage process when investigating the setting-out Dere Street and Hadrian's Wall. In other words, where the process had not been used, the author had not been able to find it.

In the case of Hadrian's Wall, it can be assumed that the two stages would have been part of the same setting-out exercise, i.e. the deviations would probably have been marked out within days of the setting-out of the long-distance alignments. In the case of Dere Street, however, the evidence indicated that at least some of the long-distance alignments may have been set out decades before the deviations were marked out from them. In fact it appeared likely that the deviations may only have been set out when the time came to construct the road, and that therefore the teams which were setting out the deviations would have been composed of quite different personnel from the team or teams which had set out the underlying long-distance alignments.

This thus raised the question which this latest monograph has sought to address: what might have been the original purpose or purposes of these long-distance alignments?

Undoubtedly, some of these alignments could have been set out simply for the purpose of establishing the overall direction that a road or structure should take. This was certainly the case with the course of Hadrian's Wall, and it is likely to have been the same for the south-western leg of the Devil's Causeway, and probably for that part of the Ribchester – Skipton road which ran through the Craven Gap. However, in other cases it is possible that the alignments may have been created for other purposes

[120] In their *Atlas of Roman Britain*, Barri Jones and David Mattingly did propose that certain Roman roads, such as the Fosse Way and the so-called *Via Devana* heading from Leicester towards Colchester, had been set out using very long-distance alignments which were at right angles to each other. Unfortunately the evidence which they cited did not support this claim. See Jones and Mattingly, 1990, 94-5 and Map 4:27. However, Rob Entwistle has recently re-examined the available evidence more scrupulously, and found that it might well appear to substantiate the case for Roman surveying of the landscapes of the Midlands and southern England on a very large scale. It will be left to Rob Entwistle to examine this possibility further and to report on it in due course.

and then parts of them re-used when the time came to set out the lines of the permanent roads across the occupied territories.

In this monograph it is suggested that these other original purposes might have included:

a) planning supply routes for the army on campaign (it is considered that much of the course of Dere Street stems from the re-use of a set of such alignments)

b) creating grids for land divisions, in the form of centuriation (for which, of course, there are records in the literature)

c) conducting rapid topographical and demographical surveys of newly-conquered territories (the course of much of the Western Main Roman Road appears to have been guided by a cluster of such alignments).

These three propositions are all postulates, but they have not been treated simply as ideas. Rather, considerable effort has been devoted to trying to ensure their practical feasibility and to identify evidence for their possible existence. As far as can be seen, all three propositions may have enjoyed an existence, in Britain, at least.

Alternative interpretations are possible, of course. With regard to c), Rob Entwistle is inclined to believe that, at times at least, such surveys might have been carried out prior to the conquest of a territory. This would certainly seem to offer a fit with some of the military expeditions described by Sherk,[121] which do appear to have been conducted with a view to prospective invasions by the Roman army at large.[122] To some extent the practice may have depended upon how heavy-handed the Roman authorities felt they could be with client kingdoms, and this would no doubt have varied according to time, situations, and personalities.

With regard to northern England, for instance, it is recorded that before the Roman conquest the Roman army had had to intervene to rescue Cartimandua, the client queen of the Brigantes, from an insurrection raised by her former consort, Venutius. It is thus at least theoretically possible that the opportunity was taken to survey substantial areas of Brigantian territory at that time. Indeed, such activities, if they had occurred, might have helped to fuel the hostilities against Cartimandua and her Roman allies. In addition, as Rob Entwistle has pointed out,[123] it is possible that the long-distance alignment underlying Roman Dere Street from Esh to the Dry Burn may have stopped at the latter point not because it had reached the limit of Roman conquest at the time (as suggested in Section 6), but because it had reached the boundary of Brigantian territory.

[121] Sherk, 1974, 537-43
[122] Austin and Rankov, 1995, 9
[123] Rob Entwistle, *pers. comm.*

In the main, though, the author is inclined to believe that the likeliest demand for such surveying activities would have come in the immediate aftermath of a period of conquest, when the hostile forces would have been subdued and the Romans would have needed to arrange for the permanent occupation and administration of the newly-acquired territory. It has been suggested that the Romans could have determined the locations of many of their forts by means of a rough set of rules, e.g. a fort every 14 to 15 miles and at each major river crossing. This would have been an entirely practical approach for a linear feature, such as a campaign trail which had been set out as a corridor across the countryside to ensure the provision of supplies and communications, but it is considered that in order to create a network of forts and roads across a wide region of newly-acquired territory – such as, for instance, large parts of northern England – a better understanding would have been required of the landscape and the peoples occupying it. Hence the need, it is considered, for such surveys.

With regard to Roman surveying practice, a factor common to all the foregoing applications, whether linear or region-wide, is that the alignments in them would have been relatively easy to set out using an instrument such as the *groma* or the *dioptra*. As proposed in Section 4 and argued in Sections 7 and 9, it is believed that this would have been coupled with the use of lights at night where longer distances had been involved.

There is one further and rather curious feature which might underpin the planning of some of the roads examined in this monograph and in BAR 492. This is the possibility that there might have been some Roman knowledge of magnetic north. Apparently, the Romans – and the Greeks before them – had been aware of the fact that if a piece of iron should be stroked repeatedly in the same direction it would become magnetised and then attract other pieces of iron to it. Currently, the general belief is that the Roman world had no knowledge of the magnetic compass. However, if the Romans had been aware of the properties of magnetism for several hundred years, it would be highly surprising if none of them had noticed that if a piece of magnetised iron were left free to move – such as when floated on a piece of wood in a bowl of water – then it would always turn to face virtually due north-south. To a surveyor, knowledge of such a property could have been a boon, especially when seeking to establish the direction of due north (or south) rapidly, or on a cloudy day.

There are some indications that at least one – or possibly a select few – Roman surveyors in Britain in the first century AD did have a knowledge of such a property. The evidence comes from three stretches of Roman alignments that appear to have been set out, as it were, into the blue. See Figure 90.

The first is the alignment used by the Western Main Roman Road, which, as described in Section 7, seems to have been set out almost due south from near Casterton to run over the Bowland Fells. This alignment appears to have been set

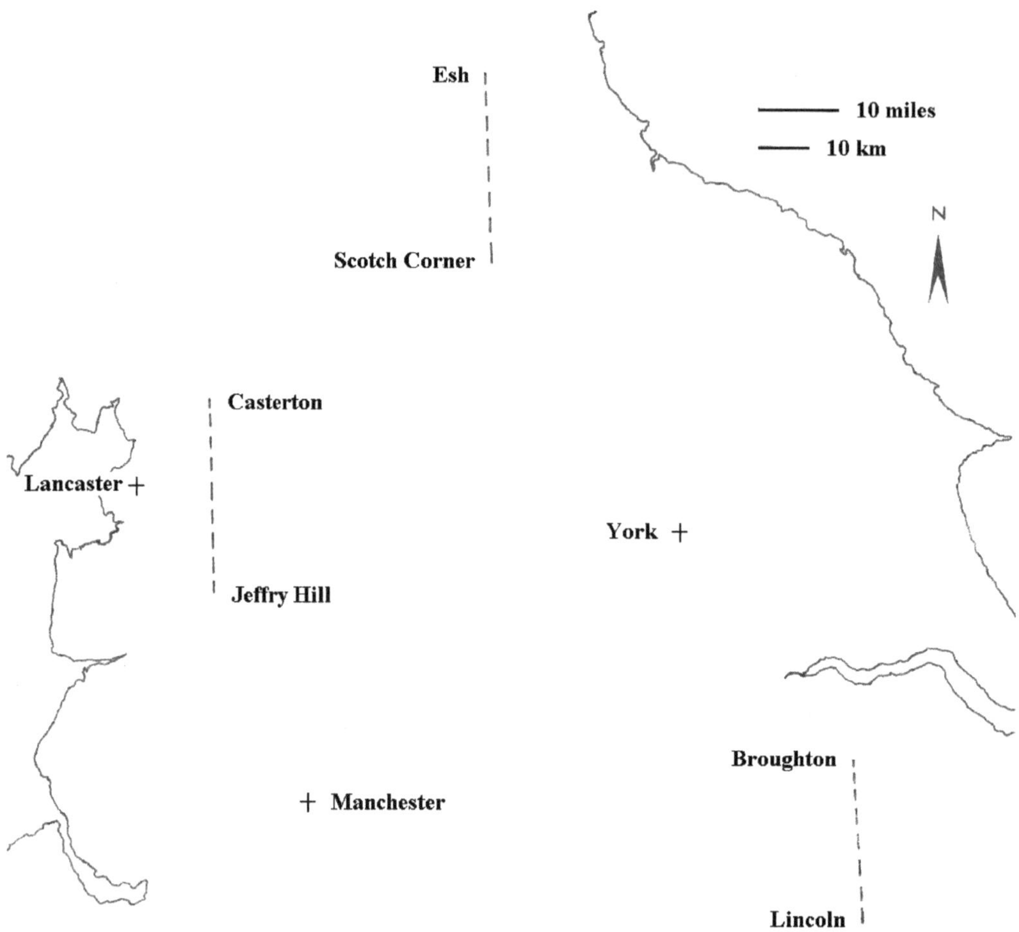

Figure 90 Three long-distance Roman alignments seemingly set out "into the blue"

out "into the blue" in order, seemingly, to find out where the line of the Roman road should be extended southwards from the Bowland Fells in order to link up with the road coming up from Manchester.

The second is the long-distance alignment which appears to underlie the course of Dere Street northwards from Scotch Corner. This appears to have been set out almost due north across the Tees Valley with no particular physical feature in sight on the skyline. As it happens, once this alignment had crossed the watershed into the valley of the River Wear, it proved to run straight up the floor of that valley, giving the planners of Dere Street considerable problems in setting out a deviation for the Roman road through Bishop Auckland, then Binchester, and then, having crossed the river, in continuing on the other side of the Wear round by Hunwick and Willington. See Figure 17. If the land north of Scotch Corner had been surveyed only in outline beforehand, much of this trouble could have been avoided by setting out a long-distance alignment only 3°or 4° to the east or west of the alignment that was chosen. It therefore appears likely that this alignment had also been set out almost due north "into the blue".

The third alignment is that of Ermine Street as it leaves Lincoln, heading almost due north for more than 20 miles (32.2 km) until it reaches a spot near Broughton from which the Humber Estuary comes into view. At this point the Roman road changes direction slightly and heads for the Roman harbour on the southern side of the Humber, at Winteringham Haven. If the Romans had surveyed the land between Lincoln and the Humber beforehand, it would have been possible for them to have set out an alignment directly from Lincoln to Winteringham. Those who know the area will appreciate that there is an escarpment to the west of the Ermine Street which falls sharply westwards towards the flatlands of the River Till and then the valley of the River Trent. However, a direct alignment from Lincoln to Winteringham would still have run to the east of this escarpment. Moreover, there is no physical feature which is visible to the north of Lincoln these days which might have prompted the Roman planners to have aimed their road upon it. It therefore appears likely that this road was likewise set out almost due north "into the blue".

To draw this evidence together, these three long-distance alignments lie on what were probably the three most important Roman roads in the north of England. Moreover, the alignments are of not insignificant lengths: the alignment over the Bowland Fells from Casterton to Jeffry Hill extends for more than 24 miles (38.9 km), whilst that from Scotch Corner to the village of Esh in County Durham is only slightly less, at 23.9 miles (38.2 km), and the alignment from Lincoln to Broughton runs for at least 20 miles (32.2 km) towards the Humber.

For the reasons explained above, all three of these alignments seem to have been set out "into the blue", and yet it can be seen that they all have a bearing which is close to 2° west of true north. For the Western Main Road over the Bowland Fells, the bearing is assessed at 1.6° west of true north. For the long-distance alignment underlying Dere Street north of Scotch Corner, the bearing is made out to be 2.1° west of true north, and for Ermine Street north of Lincoln the bearing appears to be 2.2° west of true north. Given that a piece of iron floating in a bowl of water is hardly likely to offer a stable platform from which to take a bearing, it appears quite possible that this coincidence was no accident. In the case of the Dere Street and Ermine Street alignments, too, it can be imagined that their initial courses from two of the key points in the network of Roman forts and roads – Scotch Corner and Lincoln – would have arisen from a directive for them to head due north. The surveyor setting out these bearings may have felt that no Roman commander would be likely to detect their slight divergence from true north.

Complementing these suppositions is what is understood to be the direction of magnetic north at the time. The bearing of magnetic north has varied quite considerably over the centuries. From Figure 91[124] it can be seen that between 800 and 900 BC it may have diverged as much as 20° to the east of true north. However, at the time of the Roman conquest of Britain, it appears to have lain at between 0° and 2° west of true north – in line with the long-distance alignments noted above. Thus there does seem to be a possibility that at least a small number of surveyors in Roman times may have known that a piece of magnetised iron would settle in a north-south direction if left free to turn of its own volition.

The strongest argument against the possibility that some Romans, at least, may have known about magnetic north is the complete absence of any mention of it in the Roman literature, whereas there is much on how to plot (true) north and south by other means.[125] Surely, it may be argued, it would have been impossible to have kept such an important discovery secret. In addition, as noted in Section 5, in their surveying of landscapes the Romans do not seem to have been particularly concerned about aligning their survey lines with the points of the compass. Nevertheless, perhaps the possibility there may have been some knowledge of magnetic north in Roman times is a point that should be kept in mind in case further evidence for its existence should arise at some time.

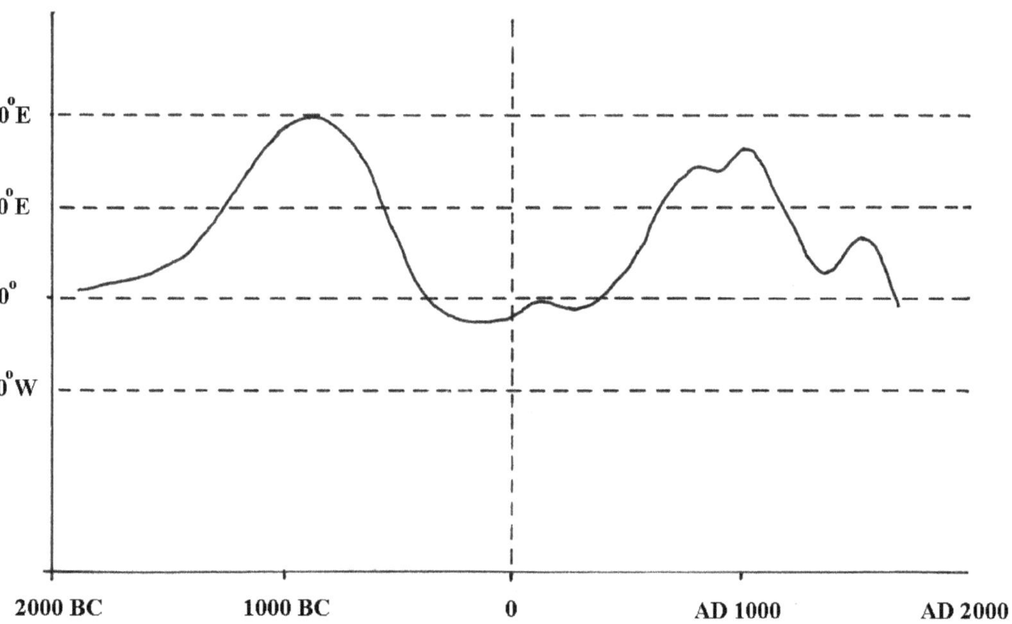

Figure 91 Estimated declination of magnetic north in Britain over the last four millennia

[124] Image extracted and redrafted from Bradford, 2007. I am indebted to Michael Lewis for this reference.
[125] Brian Campbell, *pers. comm.*

To conclude, like all the material in this monograph, the foregoing is submitted for the consideration of readers and for incorporation into the body of knowledge about Roman roads and Roman Britain. It is recognised that all of the foregoing has been derived from a narrow field of study – that of alignments, Roman roads, and their relationships to the landscapes concerned – but it is believed to have been deduced from a considerable understanding which has been gained from such studies, and that it therefore forms a solid contribution to scholarship about the Roman world.

It is almost certain that more remains to be discovered, and that there may well be a wider picture of Roman planning which remains to be deciphered yet. There are, for instance, some features about the Roman road through the Craven Gap and about the Devil's Causeway which appear not to be fully explicable by what has been understood so far. With regard to the latter, for example, if it can be accepted that the alignment of the final stretch of the Devil's Causeway towards Tweedmouth was not a coincidence, then the conclusion might appear to be inescapable that the Roman surveyors knew precisely where Tweedmouth was located, in relation to Scotch Corner. Hence it is hoped that the full publication of Rob Entwistle's discoveries will soon be accomplished, together with the more comprehensive picture that might be possible to derive from his work.

In the meantime, it may be that the alignment of the northern end of the Devil's Causeway will prove to be something of an exception. In the past many people have marvelled at the ability which the Romans apparently possessed to set out an exact alignment between two places which were not in sight of each other, and various proposals have been made about how the Roman surveyors might have been able to accomplish this. Perhaps the principal finding from this Sequel to BAR 492 is that, quite often, the explanation may have been simpler than all of these: in the main, it appears, the lines were there first.

References and sources

Austin, N. J. E. and Rankov, N. B. 1995. *Exploratio: military and political intelligence in the Roman world from the second Punic War to the Battle of Adrianople.* London, Routledge.

Bidwell, P. 2009. *The Dere Street outpost forts: Risingham and High Rochester.* Notes prepared for a field trip as part of the 2009 Pilgrimage of Hadrian's Wall (unpublished).

Bidwell, P. and Hodgson, N. 2009. *The Roman Army in Northern England.* South Shields, The Arbeia Society on behalf of the Organising Committee of the XXIst International *Limes* (Roman Frontiers Studies) Congress, Newcastle upon Tyne.

Bishop, M. C. 2005. A New Flavian Military Site at Roecliffe, North Yorkshire. *Britannia* 36, 135–223.

Bradford, University 2007. *Secular variation and calibration.* Downloaded from http://www.brad.ac.uk/archaeomagnetism/archaeomagnetic-dating/secular-variation-calibration/zananiri/ on 4 August 2013.

Breeze, D. J. 2006. *J. Collingwood Bruce's Handbook to the Roman Wall*, 14th edition, Newcastle upon Tyne, The Society of Antiquaries of Newcastle upon Tyne.

Breeze, D. J. 2011. *The Frontiers of Imperial Rome.* Barnsley, Pen & Sword Books Ltd.

Campbell, B. 2000. *The Writings of the Roman Land Surveyors: Introduction, Text, Translations, Commentary.* London, Society for the Promotion of Roman Studies.

Davies, H. E. H. 2002. *Roads in Roman Britain.* Stroud, Tempus Publishing Ltd.

Davis, J. 1853. *Report of the Secretary of War, December 1, 1853* together with *Secretary of War Jefferson Davis' instructions to the survey parties conducting the 1853 explorations and surveys for a railroad route to the Pacific.* http://cprr.org/Museum/PacRRSurvey_Secty_War_1853.html (downloaded 1 March 2012).

Della Corte, M. 1922. Groma. *Monumenti antichi: Accademia nazionale dei Lincei*, XXVIII (1922).

Della Corte, M. 1925. *Pompeii: the new excavations (houses and inhabitants).* Valle di Pompei, Tipografia di Francesco Sicignano.

Dilke, O. A. W. 1971. *The Roman land surveyors: an introduction to the agrimensores.* Newton Abbott, David and Charles.

Dilke, O. A. W. 1974. Archaeological and Epigraphic Evidence of Roman Land Surveys. *Aufstieg und Niedergang der römischen Welt*, II, 1 (1974), 564-592.

Fitzpatrick, A. P. and Scott, P. R. 1999. The Roman Bridge at Piercebridge, North Yorkshire – County Durham. *Britannia* 30, 111–32.

Frere, S. 2000. A *Limitatio* of Icenian Territory? *Britannia*, 31, 350-5.

Gates, T. and Hewitt, R. 2007. Some Newly Identified Roman Temporary Camps in Northumberland and their Relation to the Devil's Causeway. *Archaeologia Aeliana*, 5th series, 36, 11-30.

Gerhold, D. 2005. *Carriers and Coachmasters: Trade and Travel before the Turnpikes.* Chichester, Phillimore & Co. Ltd.

Graafstal, E. P. 2012. Hadrian's Haste: a priority programme for the Wall. *Archaeologia Aeliana*, 5th series, 41, 123-84.

Haken, M. 2014/5 (forthcoming). Recent Observations about the Planning of Roman Road M28 in Southern Yorkshire, and the Consequent Implications.

Hancke, T., Charlton, B. and Biggins, J. A. 2004. A geophysical survey at High Rochester Roman fort. *Archaeologia Aeliana,* 5th series, 33, 35–50.

Hodgson, N. (ed.) 2009. *Hadrian's Wall 1999-2009: A Summary of Excavation and Research prepared for the Thirteenth Pilgrimage of Hadrian's Wall, 8-14 August 2009.* Kendal, The Cumberland and Westmorland Antiquarian and Archaeological Society and The Society of Antiquaries of Newcastle upon Tyne.

Hodgson, N. 2012. *The Iron Age on the Northumberland Coastal Plain.* Tyne and Wear Archives & Museums Archaeological Monograph No. 3, Newcastle upon Tyne, TWM Archaeology and the Arbeia Society.

Jones, B. and Mattingly, D. 1990. *Atlas of Roman Britain.* Oxford, Basil Blackwood Ltd.

Lewis, M. J. T. 2001. *Surveying Instruments of Greece and Rome.* Cambridge, Cambridge University Press.

Margary, I. D. 1973. *Roman Roads in Britain.* Third edition. London, John Baker Publishers Ltd.

Moreno Gallo, I. 2004. Topografía romana. In *Elementos de ingenieria romana: congreso europeo las obras públicas romanas, 3-6 noviembre 2004, Tarragona,* Tarragona (Spain), Colegio de Ingenieros Técnicos de Obras Públicas, 2004, 25-68. (English translation by B. R. Bishop, 2006, entitled 'Roman surveying', 55p., downloaded from http://www.traianvs.net in March 2014. Note that the footnote references to this publication relate to the page numbers in this English translation.)

Ogilby, J. 1675. *Britannia* atlas: *The continuation of the Road from London to Barwick Beginning in Chester in ye Street and extending to Barwick.*

Ordnance Survey, 1996. *Forest of Bowland and Ribblesdale.* Outdoor Leisure Map 41.

Parker, M. 2010. *Locations of Ancient Woolworths Stores follow Precise Geometrical Pattern.* Press release dated 5 January 2010.

Peterson, J. 1993. *Computer-aided investigation of ancient cadastres*. http://www.uea.ac.uk/~jwmp/Peterson1993Thesis/ (downloaded 8 September 2013).

Peterson, J. W. M. 1995. *Flavian fort sites in South Wales: a spreadsheet analysis*. British Archaeological Reports International Series 600, pages 87-93, Oxford, BAR Publishing.

Peterson, J. W. M. 2006. Planned Military Landscape in Roman Britain. In *Journeys through European landscapes*, edited by L. Lévêque et al., Ponferrada (Spain), COST Action A27, 153-6. (Downloaded from http://www.uea.ac.uk/~jwmp/PMLinRB/PMLinRB.html on 20 December 2010.)

Peterson, J. 2009. *A Glossary of Agrimensorial Terms: a glossary of terms used in Roman land surveying*. http://www.uea.ac.uk/~jwmp/glossary.html#centuriation (as last updated on 13 August 2009).

Poulter, J. 1998. The date of the Stanegate, and a hypothesis about the manner and timing of the construction of Roman roads in Britain. *Archaeologia Aeliana*, 5th series, 26, 49–58.

Poulter, J. 2009. *Surveying Roman Military Landscapes across Northern Britain: the Planning of Roman Dere Street, Hadrian's Wall and the Vallum, and the Antonine Wall in Scotland* with a supplementary report *Mapping Matters with the Antonine Wall* by Peter McKeague. British Archaeological Reports 492, Oxford, BAR Publishing.

Poulter, J. 2010. *The Planning of Roman Roads and Walls in Northern Britain*. Chalford (Stroud), Amberley Publishing.

Ramm, H. 1980. Native settlements East of the Pennines. In *Rome and the Brigantes*, edited by Keith Branigan, Sheffield, University of Sheffield, 28-40.

Ross, P. 1921. Roman Roads in Yorkshire, Ribchester to York. *The Bradford Antiquary*, vol. 6 (new series, vol. 4), 33-64. Note: in the same volume, P. Ross describes The First Stage of the Roman Road from Ribchester to York (The Road between Downham Park and Ribchester) on pages 267-72.

Roy, W. 1790. An Account of the Trigonometrical Operation, whereby the Distance between the Meridians of the Royal Observatories of Greenwich and Paris has been determined. *Philosophical Transactions of the Royal Society of London*, 80 (1790), 111-614.

Schiöler, T. 1994. The Pompeii-*groma* in new light. *Analecta Romana*, 22 (1994), 45-60.

Sherk, R. K. 1974. Roman Geographical Exploration and Military Maps. *Aufstieg und Niedergang der römischen Welt*, II, 1 (1974), 534-62.

Snape, M. 2014. The Roman bridge at Corbridge: its setting and landscape history. *Arbeia Journal*, 10 (2014).

Taylor, G. and Skinner, A. 1776. *Survey and maps of the roads of North Britain or Scotland*. London, published by Act of Parliament.

Toller, H. 2014 (forthcoming). The Roman road between Ewe Close, Crosby Ravensworth and Kirkby Thore. *Transactions of the Cumberland and Westmorland Antiquarian and Archaeological Society*, 3rd series, XIV (2014).

Wacher, J. 1975. *The Towns of Roman Britain*. London, B. T. Batsford Ltd.

Went, D. and Ainsworth, S. 2013. Whitley Castle, Northumberland: An Analytical Survey of the Fort and its Setting. *Britannia*, 44, 93-143.

Williamson, T. and Bellamy, L. 1983. *Ley lines in question*. Tadworth, Surrey, World's Work Ltd.

Woolliscroft, D. J. and Hoffmann, B. 2006. *Rome's First Frontier: The Flavian Occupation of Northern Scotland*. Stroud, Tempus Publishing Ltd.

www.ingramcontent.com/pod-product-compliance
Lightning Source LLC
LaVergne TN
LVHW071220210926
848482LV00015B/858